# Mixed Use and Residential Tenants' Rights:
## The Landlord and Tenant Act 1987 and Leasehold Enfranchisement

Peta Dollar and Sarah Thompson-Copsey

AMSTERDAM • BOSTON • HEIDELBERG • LONDON • NEW YORK
OXFORD • PARIS • SAN DIEGO • SAN FRANCISCO • SINGAPORE

21 0336292 3

344.10643 DOL

EG Books is an imprint of Elsevier
The Boulevard, Langford Lane, Kidlington, Oxford, OX5 1GB, UK
30 Corporate Drive, Suite 400, Burlington, MA 01803, USA

First edition 2010
Copyright © 2010, Peta Dollar and Sarah Thompson-Copsey
(except in relation to extracts from statute, where Crown copyright applies)
Published by Elsevier Ltd. All rights reserved.

The right of Peta Dollar and Sarah Thompson-Copsey to be identified as the authors of this work has been asserted in accordance with the Copyright, Designs and Patents Act 1988

No part of this publication may be reproduced or transmitted in any form or by any means, electronic or mechanical, including photocopying, recording, or any information storage and retrieval system, without permission in writing from the publisher. Details on how to seek permission, further information about the Publisher's permissions policies and our arrangements with organizations such as the Copyright Clearance Center and the Copyright Licensing Agency, can be found at our website: www.elsevier.com/permissions.

This book and the individual contributions contained in it are protected under copyright by the Publisher (other than as may be noted herein).

**Notices**
Knowledge and best practice in this field are constantly changing. As new research and experience broaden our understanding, changes in research methods, professional practices, or medical treatment may become necessary.

Practitioners and researchers must always rely on their own experience and knowledge in evaluating and using any information, methods, compounds, or experiments described herein. In using such information or methods they should be mindful of their own safety and the safety of others, including parties for whom they have a professional responsibility.

To the fullest extent of the law, neither the Publisher nor the authors, contributors, or editors, assume any liability for any injury and/or damage to persons or property as a matter of products liability, negligence or otherwise, or from any use of operation of any methods, products, instructions, or ideas contained in the material herein.

**British Library Cataloguing-in-Publication Data**
A catalogue record for this book is available from the British Library

**Library of Congress Cataloging-in-Publication Data**
A catalog record for this book is available from the Library of Congress

ISBN 978-07282-0571-0

For information on all EG Books publications visit our website at elsevierdirect.com

Typeset in Palatino 10/12 by Amy Boyle
Cover design by Rebecca Caro
Printed and bound in Great Britain

09 10 11 12   10 9 8 7 6 5 4 3 2 1

Working together to grow
libraries in developing countries

www.elsevier.com | www.bookaid.org | www.sabre.org

ELSEVIER   BOOK AID International   Sabre Foundation

# Contents

Foreword and acknowledgements .............................. v

Table of Cases ............................................... vii
Table of Statutes ............................................. xi
Table of Statutory Instruments ............................... xix

| 1 | Introduction to Mixed Use Buildings ..................... 1 |
|---|---|
| 2 | Outline of the Impact of Residential Tenants' Rights on Mixed Use Properties ................................ 3 |
|   | A. The Landlord and Tenant Act 1987, Part I (as amended)—tenants' rights of first refusal ............ 4 |
|   | B. The Leasehold Reform Act 1967 (as amended) —tenant's enfranchisement rights (houses) ............. 6 |
|   | C. The Leasehold Reform, Housing and Urban Development Act 1993 (as amended) —collective enfranchisement (flats) ..................... 8 |
|   | D. The Commonhold and Leasehold Reform Act 2002 —the right to manage ............................... 9 |
| 3 | The Landlord and Tenant Act 1987, Part I (as amended)—Tenants' Rights of First Refusal ........... 11 |
|   | A. Determining whether or not the 1987 Act applies ....... 11 |
|   | B. If the 1987 Act does apply to the intended disposal, what must the landlord do before making the disposal? ........................................ 16 |
|   | C. What happens then? ................................. 18 |
|   | D. After acceptance and nomination .................... 19 |

    E. Exercising rights against a subsequent purchaser
       (where the 1987 Act has not been complied with) . . . . . . . 23
       Flowchart A: Must the landlord comply with the
       1987 Act in making a proposed disposal? . . . . . . . . . . . . . . 25
       Flowchart B: Have the tenants protected their rights,
       or is the landlord free to sell the building? . . . . . . . . . . . . . 26
       Flowchart C: Costs and other consequences of service
       of landlord's notice of withdrawal under section 9B . . . . . 28
       Flowchart D: Costs and other consequences of
       of nominated person's notice of withdrawal
       under section 9A. . . . . . . . . . . . . . . . . . . . . . . . . . . . . . . . . . . 29
       Flowchart E: Tenants' rights following unauthorised
       disposal by landlord . . . . . . . . . . . . . . . . . . . . . . . . . . . . . . . 30

4   **The Landlord and Tenant Act 1987, Part I (as amended):
    Section by Section**. . . . . . . . . . . . . . . . . . . . . . . . . . . . . . . . . . . . 31

5   **Collective Enfranchisement Rights and Rights
    to Manage** . . . . . . . . . . . . . . . . . . . . . . . . . . . . . . . . . . . . . . . . . . 191
    A. The Leasehold Reform, Housing and Urban
       Development Act 1993 (as amended)—collective
       enfranchisement (flats) . . . . . . . . . . . . . . . . . . . . . . . . . . . . 191
       Flowchart F: Collective enfranchisement: timetable
       and consequences. . . . . . . . . . . . . . . . . . . . . . . . . . . . . . . . 209
    B. The Commonhold and Leasehold Reform Act 2002
       —the right to manage . . . . . . . . . . . . . . . . . . . . . . . . . . . . 210
       Flowchart G: Right to manage timetable . . . . . . . . . . . . . . 218

6   **Structuring to Avoid the Need for Compliance with
    the Acts** . . . . . . . . . . . . . . . . . . . . . . . . . . . . . . . . . . . . . . . . . . . 219
    A. The Landlord and Tenant Act 1987 . . . . . . . . . . . . . . . . . . 219
    B. The Leasehold Reform, Housing and Urban
       Development Act 1993 and the Commonhold and
       Leasehold Reform Act 2002 . . . . . . . . . . . . . . . . . . . . . . . . 225
    C. The 1993 Act and the estate management scheme . . . . . . 227

**Appendices**
A.  The Leasehold Reform, Housing and Urban Development
    Act 1993, Part 1, Chapter 1 (as amended). . . . . . . . . . . . . . . . 229
B.  The Commonhold and Leasehold Reform Act 2002,
    Part II, Chapter 1 . . . . . . . . . . . . . . . . . . . . . . . . . . . . . . . . . . . 294
Index . . . . . . . . . . . . . . . . . . . . . . . . . . . . . . . . . . . . . . . . . . . . . . . . 331

# Foreword and Acknowledgements

We were first asked to provide training on mixed use buildings and the implications of residential tenants' rights back in 2005, mainly as a result of our wide experience of the Landlord and Tenant Act 1987, initially arising out of the sale of the South Kensington Estate by The Trustees of Henry Smith's Charity to The Wellcome Trust in the mid 1990s (then the largest ever residential property sale in this country), as well as subsequent transactions. Since then, we have provided this training to many people up and down the country, and it is out of this training that this book has come.

Our thanks to Colin Davey, Director of Professional Development at the College of Law, who originally asked us to prepare a course on this topic, Susan Highmore, Katharine Fenn and Kay Balaam, who helped with the initial preparation of the course, and to all those who participated in our classes and lectures. Our thanks also to Anthony Radevsky, who first came up with the argument that a partly constructed development is not a "building" for the purposes of the 1987 Act; this argument was subsequently developed in an article that we wrote jointly with him called "Developer Beware! The pitfalls of tenants' rights of first refusal and collective enfranchisement" and published in *The Landlord and Tenant Review*, Volume 10, Issue 2 (March/April 2006).

We would also like to thank all those at EG Books who have been involved in the production of this book, including Sarah Jackman, Alison Bird, Victoria Broom, Paul Sayers, Adrian Gilbert and Amy Boyle.

But this book is not confined to landlords and commercial occupiers seeking to understand the rights of residential tenants. We also focus on the rights of those tenants and what they need to do to protect and exercise their rights. This legislation is tricky and it is easy to get things wrong. We hope that this practical guide will help all those involved with mixed use buildings, whether as owners, tenants, developers or advisors, to have a better understanding of the rights conferred by this legislation and to avoid the pitfalls that it contains.

Our intention is that those to whom this legislation is new should start by reading Chapter 2, which will give them a brief general outline of the 1987 Act (tenants' rights of first refusal), the 1993 Act (tenants' rights of collective enfranchisement) and the 2002 Act (tenants' collective rights to manage), as well as a brief look at the Leasehold Reform Act 1967 (as amended) (the right of an individual tenant of a "house" to enfranchise) in the context of mixed use buildings. They may then wish to pass on to Chapter 3, which contains a more detailed analysis of the 1987 Act and Chapter 5, which contains a more detailed analysis of the 1993 and 2002 Acts.

Chapter 4 is the heart of this book. It sets out Part 1 of the 1987 Act in full and contains an in-depth section-by-section examination of the provisions of the 1987 Act. It is likely to be of most use to lawyers and other advisors dealing with those provisions, whether on behalf of landlords or tenants, and incorporates the individual sections of the 1987 Act. Chapter 6 deals with possible ways of structuring so as to avoid the need for compliance with the legislation, and is aimed primarily at landlords and their advisors. Finally, the Appendices contain the relevant provisions of the 1993 and the 2002 Acts.

References in this book to "the 1987 Act" refer to Part 1 of the Landlord and Tenant Act 1987 as amended by the Housing Act 1996.

References in this book to "the 1993 Act" refer to Chapter 1 of Part 1 of the Leasehold Reform, Housing and Urban Development Act 1993 as amended by the Commonhold and Leasehold Reform Act 2002.

References in this book to "the 2002 Act" refer to Chapter 1 of Part 2 of the Commonhold and Leasehold Reform Act 2002.

The law is stated as at 30 April 2009.

**Peta Dollar**
LLB (Hons.), Non-Practising Solicitor
**Sarah Thompson-Copsey**
BA (Hons.), LLM, Non-Practising Solicitor

*April 2009*

# Table of Cases

7 Strathray Gardens Ltd *v* Pointstar Shipping and Finance
    Ltd [2004] EWCA Civ 1669; [2005] 01 EG 95 (CS) ...... Para 5-38
9 Cornwall Crescent London Ltd *v* Royal Borough of
    Kensington & Chelsea [2005] EWCA Civ 324.......... Para 5-37

Bearmans Ltd *v* Metropolitan Police Receiver [1961]
    1 WLR 634....................................... Para 4-484
Belvedere Court Management Ltd *v* Frogmore Developments
    Ltd [1996] 1 EGLR 59 ............................ Para 4-362
Boss Holdings Ltd *v* Grosvenor West End Properties [2008]
    UKHL 5........................................ Para 2-13/17
Boyle *v* Hallstate Ltd [2002] EWHC 972 (CH)............ Para 4-362,
    ................................................ 4-452, 4-454
Brick Farm Management Ltd *v* Richmond Housing Trust Ltd
    [2005] EWHC 1650 ................................ Para 5-18
Brown & Root Ltd *v* Sun Alliance and London Assurance
    Company Ltd [1996] EWCA Civ 1261 ............... Para 4-490
Burman *v* Mount Cook Land Ltd [2001] EWCA Civ 1712;
    [2001] 48 EG 128 (CS) ............................ Para 5-39

Cadogan *v* Morris [1999] 1 EGLR 59 .................... Para 5-27
Cadogan *v* Sportelli [2008] UKHL 71 ................... Para 5-59
Cawthorne *v* Hamdan [2007] EWCA Civ 6................ Para 5-44
Chinnock *v* Hocaoglu [2007] EWHC 2933 (Ch.);
    [2008] 2 EGLR 77 ................................ Para 4-96

Dartmouth Court Blackheath Ltd *v* Berisworth Ltd [2008]
    EWHC 350 (Ch)................ Para 4-04, 4-17, 4-47, 4-65, 4-309
Denetower Ltd *v* Toop [1991] EGCS 39 .............. Para 4-17, 4-65

Englefield Court Tenants v Skeels [1990] 2 EGLR 230......Para 4-362

Gaingold Ltd v WHRA RTM Company [2006]
 03 EG 122 ................................Para 4-23, 5-12
Green v Westleigh Properties Ltd [2008] EWHC
 1474 (QB)....................... Para 4-83, 4-294, 4-307, 4-367
Gregory v Saddiq [1991] 1 EGLR 237....................Para 4-362
Grosvenor Estates Ltd v Prospect Estates Ltd [2008]
 EWCA Civ 1281................................Para 2-16/17

Henry Smith's Charity Trustees v Kyriacou [2001]
 Ch 493.........................................Para 4-178,

James v United Kingdom (1968) 8 ECHR 123..............Para 2-10

Kay-Green v Twinsectra Ltd [1996] 2 EGLR 43 ............Para 4-17
Kensington Heights Commercial Company Ltd v Campden
 Hill Developments Ltd [2007] EWCA Civ 245;
 [2007] 1 EGLR 130.......... Para 4-320, 4-342, 4-349, 4-376, 4-396

Lay v Ackerman [2004] EWCA Civ 184...................Para 5-39
London Rent Assessment Panel v Holding and Management
 (Solitaire) Ltd [2007] EW Lands LRX/138/2006........Para 5-11
Long Acre Securities Ltd v Karet [2004] EWHC 442 ........Para 4-17

M25 Group Ltd v Tudor [2003] EWCA Civ 1760...... Para 4-83, 4-307
Mainwaring v Henry Smith's Charity Trustees [1996]
 2 EGLR 25 ....................... Para 4-47, 4-79, 4-97, 4-307
Mainwaring v Henry Smith's Charity (No 2) [1996]
 EWCA Civ 657 ............................ Para 4-130, 4-133
Mannai Investments v Eagle Star Life Assurance [1997] AC 749;
 [1997] EGCS 82............................. Para 4-83, 4-307
Marine Court (St Leonards on Sea) Freeholders Ltd v Rother
 District Investments Ltd [2008] 02 EG 148 ............Para 5-13
Michaels v Harley House [1998] EWCA Civ 1714;
 [1998] EGCS 159 ...........................Para 4-307, 6-09
Michaels v Taylor Woodrow Developments Ltd [2000]
 PLSCS 101........................................Para 6-09
Mohammed El Naschie v The Pitt Place (Epsom) Ltd
 [1998] NPC 83 ..................................Para 4-133
Moir v Williams [1892] 1 QB 264........................Para 4-17

National Westminster Bank plc *v* Jones [2000] EG 82 (CS)... Para 6-07

Oakwood Court (Holland Park) Ltd *v* Daejan Properties
   Ltd [2007] 1 EGLR 121 ............................. Para 5-11
Okonedo *v* Kirby [2006] EWLands LRX_15_2006 .......... Para 4-47

Saga Properties Ltd *v* Palmeira Square Nos 2-6 Ltd [1995]
   15 EG 109 ........................................ Para 4-17
Savva *v* Philip Galway-Cooper [2005] EWCA Civ 1068;
   [2005] 28 EG 120 (CS) ............ Para 4-307, 4-330, 4-353, 4-381
Scmlla Properties Ltd *v* Gesso Properties (BVI) Ltd [1995]
   EGCS52 ........................................... Para 4-29
Sheffield District Railway *v* Great Central Railway Company
   [1911] 27 TLR 451................................. Para 4-213
Sinclair Gardens Investments (Kensington) Ltd *v* Poets Chase
   Freehold Company Ltd [2007] EWHC 1776 (Ch) ...... Para 5-28
Spiro *v* Glencrown Properties Ltd [1991] 1 EGLR 185...... Para 4-320
Staszewski *v* Maribella Ltd [1998] 1 EGLR 34 ............ Para 4-307

Tandon *v* Trustees of Spurgeons Homes [1982] 2 EGLR 73.. Para 2-12

Willingale *v* Globalrange Ltd [2000] 2 EGLR 55 ....... Para 5-27, 5-40
Woodridge Ltd *v* Downie [1997] 2 EGLR 193.............. Para 4-97

# Table of Statutes

Charities Act 1993 .................................... Para 4-60
Civil Justice Act 1982
    s 37(2)........................................ Para 4-286
Commonhold and Leasehold Reform Act 2002 ....... Para 2-02, 2-11,
                   2-18, 2-21/23, 5-02, 5-05, 5-61/5-84, 6-13/15
    s 73 .......................................... Para 5-66
    s 74 .......................................... Para 5-66
    s 82 .......................................... Para 5-77
    s 92 .......................................... Para 5-80
Companies Act 1985 ...........................................
    s 736 ............................... Para 4-37, 4-47, 4-62

Housing Act 1974..................................... Para 2-11
Housing Act 1980
    s 52 .......................................... Para 4-37
Housing Act 1985
    s 1 ........................................... Para 4-290
    s 5(4).......................................... Para 4-29
    s 5(5).......................................... Para 4-29
    s 6 ........................................... Para 4-29
Housing Act 1988, Part III ............................ Para 4-29
Housing Act 1996 .... Para 2-04, 2-11, 3-18, 4-47, 4-79, 4-97, 4-133, 6-09
Housing Associations Act 1985
    s 1(1).......................................... Para 4-29
    s 1(2).......................................... Para 4-29

Interpretation Act 1978
    s 6(c).......................................... Para 4-133

Landlord and Tenant Act 1954 . . . . . . . . . . . . . . . . . . . . . Para 3-09, 3-11
    Part I . . . . . . . . . . . . . . . . . . . . . . . . . . . . . . . . . . . . . . . . . . . Para 5-16
    Part II . . . . . . . . . . . . . . . . . . . . . . . . . . . . . Para 4-37, 4-47, 5-18
    s 38(4)(a) . . . . . . . . . . . . . . . . . . . . . . . . . . . . . . . . . . . . . . Para 4-37
Landlord and Tenant Act 1985
    s 3 . . . . . . . . . . . . . . . . . . . . . . . . . . . . . . . . . . . . . . Para 4-314, 4-315
    s 3A . . . . . . . . . . . . . . . . . . . . . Para 3-36, 4-307, 4-314, 4-315, 4-352
    s 25 . . . . . . . . . . . . . . . . . . . . . . . . . . . . . . . . . . . . . . . . . . Para 5-77
Landlord and Tenant Act 1987 . . . . . . . . . . Para 2-02/07, 2-21, 3-01/37,
         . . . . . . . . . . . 4-01/491, 5-01, 5-32, 5-61, 6-01/6-12, 6-13, 6-14, 6-15
    s 1(1) . . . . . . . . . . . . . . . . . . . . . Para 3-01, 4-01/15, 4-17, 4-79, 4-94
    s 1(2) . . . . . . . . . . . . . . . . . . . . . . . . . . . . . . . . . . . Para 4-05, 4-16/21
    s 1(3) . . . . . . . . . . . . . . . . . . . . . . . . . . . . . . . . . . . . . . . . Para 4-22/23
    s 1(4) . . . . . . . . . . . . . . . . . . . . . . . . . . . . Para 4-24/25, 4-29, 4-55
    s 1(5) . . . . . . . . . . . . . . . . . . . . . . . . . . . . . . . . . . . Para 4-26/27, 4-491
    s 2(1) . . . . . . . . . . . . . . . . . . . . . . . . . . . . . . . . . . . . . . . . Para 4-28/32
    s 2(2) . . . . . . . . . . . . . . . . . . . . . . . . . . . . . . . . . Para 4-29, 4-33/34
    s 3 . . . . . . . . . . . . . . . . . . . . . . . . . . . . . . . . . . . . . . . . . . . Para 4-37
    s 3(1) . . . . . . . . . . . . . . . . . . . . . . . . . . . . . . . . . . . . . . . . Para 4-35/39
    s 3(2) . . . . . . . . . . . . . . . . . . . . . . . . . . . . . . . . . Para 4-37, 4-40/41
    s 3(3) . . . . . . . . . . . . . . . . . . . . . . . . . . . . . . . . . Para 4-37, 4-42/43
    s 3(4) . . . . . . . . . . . . . . . . . . . . . . . . . . . . . . . . . Para 4-37, 4-44/45
    s 4(1) . . . . . . . . . . . . . . . . . . . . . . . . . . . . . . . . . Para 4-17, 4-46/52
    s 4(1A) . . . . . . . . . . . . . . . . . . . . . . . . . . . Para 4-29, 4-47, 4-53/55
    s 4(2) . . . . . . . . . . . . . . . . . . . Para 4-47, 4-54/55, 4-56/62, 4-362
    s 4(3) . . . . . . . . . . . . . . . . . . . . . . . . . . . . . . . . . . . . Para 4-47, 4-63
    s 4(4) . . . . . . . . . . . . . . . . . . . . . . . . . . . . . . . . . . Para 4-47, 4-64/68
    s 4(5) . . . . . . . . . . . . . . . . . . . . . . . . . . . . . . . . . . . . . . . . . . Para 4-69
    s 4(6) . . . . . . . . . . . . . . . . . . . . . . . . . . . . . . . . . . . . . . . . . . Para 4-70
    s 4A(1) . . . . . . . . . . . . . . . . . . . . . . . . . . . . . . . . . Para 4-47, 4-71/72
    s 4A(2) . . . . . . . . . . . . . . . . . . . . . . . . . . . . Para 4-47, 4-54, 4-73/74
    s 4A(3) . . . . . . . . . . . . . . . . . . . . . . . . . . . . . . . . Para 4-47, 4-74, 4-75
    s 4A(4) . . . . . . . . . . . . . . . . . . . . . . . . . . . . Para 4-47, 4-74, 4-76/77
    s 5 . . . . . . . . . Para 3-01, 3-12, 3-16, 3-18/20, 3-35, 4-14, 4-15, 4-47,
    . . . 4-74, 4-79, 4-81, 4-83, 4-88, 4-92, 4-98, 4-99, 4-124, 4-125, 4-294,
    . . . . . . . . . . . . . . . . . . . . . . . . . . . . . 4-307, 4-320, 4-389, 4-461, 4-477
    s 5(1) . . . . . . . . . . . . . . . . . . Para 4-08, 4-78/86, 4-100, 4-125, 4-309
    s 5(2) . . . . . . . . . . . . . . . . . . . . . . . . . . . . . . . . . . . . . . . . Para 4-87/88
    s 5(3) . . . . . . . . . . . . . . . . . . . . . . . . . . . . Para 4-89/90, 4-340, 4-386
    s 5(4) . . . . . . . . . . . . . . . . . . . . . . . . . . . . . . . . . . . . . . . . Para 4-91/92
    s 5(5) . . . . . . . . . . . . . . . . . . . . . . . . . . . . . Para 4-14, 4-92, 4-93/94

*Table of Statutes*

s 5A .... Para 3-12, 3-14, 4-81, 4-88, 4-95/101, 4-120, 4-204, 4-320
s 5A(4) ..................................... Para 4-83
s 5A(5)................................... Para 4-83, 4-135
s 5B........... Para 3-12, 3-15, 3-19, 4-79, 4-81, 4-88, 4-102/109,
................. 4-120, 4-132, 4-135, 4-154, 4-204, 4-320
s 5B(2)..................................... Para 4-105
s 5B(5)...................................... Para 4-83
s 5B(6) ................................. Para 4-83, 4-135
s 5B(7)...................................... Para 4-79
s 5C.............. Para 3-12, 3-15, 4-81, 4-110/114, 4-120, 4-204
s 5C(2)..................................... Para 4-112
s 5C(4) ..................................... Para 4-83
s 5C(5)................................... Para 4-83, 4-135
s 5D.............. Para 3-12, 3-15, 4-81, 4-115/118, 4-120, 4-204
s 5D(2) .................................... Para 4-117
s 5D(4) ..................................... Para 4-83
s 5D(5)................................... Para 4-83, 4-135
s 5E .................. Para 3-12, 3-15, 4-81, 4-119/122, 4-204
s 5E(3)...................................... Para 4-83
ss 6-10........................ Para 3-01, 4-15, 4-282, 4-294
s 6........................................ Para 3-14, 4-15
s 6(1)..................................... Para 4-123/130
s 6(2)..................................... Para 4-131/138
s 6(3) ........................... Para 4-133, 4-134, 4-139/140
s 6(4)............................... Para 4-135, 4-139/140
s 6(5) ......................... Para 4-130, 4-137/140, 4-204
s 6(6) .............................. Para 4-141/142, 4-391
s 6(7).............................. Para 4-143/144, 4-393
s 7(1)..................................... Para 4-145/156
s 7(2) ........................ Para 4-146, 4-156, 4-157/158
s 7(3) ........................ Para 4-146, 4-156, 4-157/158
s 7(4).................................... Para 44-157/158
s 8(1).................................... Para 4-159/167
s 8(2).................................... Para 4-168/170
s 8(3) ......................... Para 4-137, 4-171/172, 4-174
s 8(4) ....................... Para 4-172, 4-173/174, 4-182
s 8(5)..................................... Para 4-175
s 8A....................... Para 3-25, 3-26/28, 4-160, 4-177
s 8A(1) .................................. Para 4-176/178
s 8A(2) .................................. Para 4-179/180
s 8A(3) .................................. Para 4-179/180

s 8A(4) . . . . . . . . . . . . . . . . . . . . . . . . . . . . . . . . . . . . Para 4-181/182
s 8A(5) . . . . . . . . . . . . . . . . . . . . . . . . . . . . . . . . . . . . Para 4-181/182
s 8A(6) . . . . . . . . . . . . . . . . . . . . . . . . . . . . . . Para 4-183/4, 4-449
s 8B . . . . . . . . . . . . . . . . . . . . . . Para 3-25, 3-29/30, 4-160, 4-186
s 8B(1) . . . . . . . . . . . . . . . . . . . . . . . . . . . . . . . . . . . . Para 4-185/187
s 8B(2) . . . . . . . . . . . . . . . . . . . . . . . . . Para 4-188/192, 4-201, 4-202
s 8B(3) . . . . . . . . . . . . . . . . . . . . . . . . . . . . . Para 4-193/194, 4-199
s 8B(4) . . . . . . . . Para 4-195/197, 4-199, 4-200, 4-201, 4-202, 4-248
s 8B(5) . . . . . . . . . . . . . . . . . . . . . . . . . . . . . . . . . . . . Para 4-198/200
s 8B(6) . . . . . . . . . . . . . . . . . . . . . . . . . . . . . . . Para 4-190, 4-201/202
s 8C . . . . . . . . . . . . . . . . . . . . . . . . Para 4-120, 4-160, 4-203/208
s 8C(4) . . . . . . . . . . . . . . . . . . . . . . . . . . . . . . . . . Para 4-399, 4-401
s 8D . . . . . . . . . . . . . . . . . . . . . . Para 3-11, 4-47, 4-160, 4-210
s 8D(1) . . . . . . . . . . . . . . . . . . . . . . . . . . . . . . . . . . . Para 4-209/210
s 8D(2) . . . . . . . . . . . . . . . . . . . . . . . . . . . . . . . . . . . . . . . Para 4-211
s 8E(1) . . . . . . . . . . . . . . . . . . . . . . . . . . . . . Para 4-212/213, 4-215
s 8E(2) . . . . . . . . . . . . . . . . . . . . . . . . . . . . . . . . . . Para 4-214/215
s 8E(3) . . . . . . . . . . . . . . . . . . . . . . . . . . . . Para 4-213, 4-216/217
s 8E(4) . . . . . . . . . . . . . . . . . . . . . . . . . . . . . Para 4-213, 4-218/219
s 8E(5) . . . . . . . . . . . . . . . . . . . . . . . . . . . . . Para 4-213, 4-218/219
s 8E(6) . . . . . . . . . . . . . . . . . . . . . . . . . . . . . Para 4-213, 4-220/224
s 9A . . . . . . . . . . . . . . . . . . . . . . . Para 4-182, 4-201, 4-215, 4-414
s 9A(1) . . . . . . . . . . . . . . . . . . . . . . . . . Para 4-225/231, 4-233, 4-235
s 9A(2) . . . . . . . . . . . . . . . . . . . . . . . . Para 4-133, 4-140, 4-232/235
s 9A(3) . . . . . . . . . . . . . . . . . . . . . . . . . . . . . . . . . . Para 4-236/237
s 9A(4) . . . . . . . . . . . . . . . . . . . . . . . . . . . Para 4-237, 4-238/239
s 9A(5) . . . . . . . . . . . . . . . . . . . . . . . . . . . . Para 4-237, 4-238/239
s 9A(6) . . . . . . . . . . . . . . . . . . . . . . . . . . . . . . . . . . . Para 4-240/245
s 9A(7) . . . . . . . . . . . . . . . . . . . . . . . . . . . . . . . . . . . Para 4-240/245
s 9A(8) . . . . . . . . . . . . . . . . . . . . . . . . . . . . . . . . . . . Para 4-246/248
s 9B . . . . . . . . . . . . . . . . . . . . . . . . . . . . . . . . . Para 4-180, 4-215
s 9B(1) . . . . . . . . . . . . . . . . . . . . . . . . . . . . . Para 4-249/255, 4-258
s 9B(2) . . . . . . . . . . . . . . . . . . . . . . . . . . . . . . . . . . . Para 4-256/258
s 9B(3) . . . . . . . . . . . . . . . . . . . . . . . . . . . . . . . . . . . Para 4-259/262
s 9B(4) . . . . . . . . . . . . . . . . . . . . . . . . . . . . . . . . . . . Para 4-259/262
s 9B(5) . . . . . . . . . . . . . . . . . . . . . . . . . . . . . . . . . . . Para 4-263/265
s 10 . . . . . . . . . . . . . . . . . . . . . . . . . . . . . . . . . . . . . . . . Para 4-215
s 10(1) . . . . . . . . . . . . . . Para 4-266/271, 4-273, 4-275, 4-437, 4-457
s 10(2) . . . . . . . . . . . . . . . . . . . . . . . . . . . . . . . Para 4-272/3, 4-447
s 10(3) . . . . . . . . . . . . . . . . . . . . . . . . Para 4-267, 4-274/5, 4-437
s 10(4) . . . . . . . . . . . . . . . . . . . . . . . . . . . . . . . . . . . Para 4-276/278

*Table of Statutes*

s 10(5) . . . . . . . . . . . . . . . . . . . . Para 4-247, 4-264, 4-277, 4-279/280
s 10A(1) . . . . . . . . . . . . . . . . . . . . Para 4-08, 4-79, 4-204, 4-281/284
s 10A(2) . . . . . . . . . . . . . . . . . . . . . . . . . . . . . . . . . . . . . Para 4-285/286
s 10A(3) . . . . . . . . . . . . . . . . . . . . . . . . . . . . Para 4-282, 4-287/288
s 10A(4) . . . . . . . . . . . . . . . . . . . . . . . . . . . . . . . . . . . . . Para 4-289/290
s 10A(5) . . . . . . . . . . . . . . . . . . . . . . . . . . . . . . . . . . . . . Para 4-291/292
ss 11-17 . . . . . . . . . . . . Para 4-120, 4-204, 4-282, 4-292, 4-340, 4-386
s 11(1) . . . . . . . . . . . . . . . . . . . . . . . . . . . . . . . . Para 4-293/299, 4-305
s 11(2) . . . . . . . . . . . . . . . . . . . . Para 4-100, 4-294, 4-300/303, 4-309
s 11(3) . . . . . . . . . . . . . . . . . . . . . Para 4-47, 4-304/305, 4-312, 4-466
s 11(A) . . . . . . . . . Para 4-98, 4-306/316, 4-318, 4-329, 4-342, 4-352,
. . . . . . . . . . . . . . . . . . . . . . . . . . . . . 4-369, 4-380, 4-419, 4-433, 4-438
s 11A(2) . . . . . . . . . . . . . . . . . . . . . . . . . . . . . . . . . . . . . . . . . Para 4-426
s 11A(4) . . . . . . . . . . . . . . . . . . . . . . . . . . . . . . . . . . . . . . . . . Para 4-423
s 12 . . . . . . . . . . . . . . . . . . . . . . . . . . . . . . . . . . . . . . Para 4-307, 4-362
ss 12A-14 . . . . . . . . . . . . . . . . . . . . . . . . . . . . . . . . Para 4-433, 4-435
s 12A . . . . . . . . . Para 4-98, 4-307, 4-312, 4-321, 4-333, 4-342, 4-345,
. . . . . . . . . . . . . . . . . . . . . . . . . . . . . . 4-349, 4-389, 4-430, 4-433, 4-449
s 12A(1) . . . . . . . Para 4-317/327, 4-334, 4-335, 4-337, 4-338, 4-342
s 12A(2) . . . . . . . . . . . . . . . . . . . . . . . . . . . . . . . Para 4-328/331, 4-342
s 12A(3) . . . . . . . . . . . . . . . . . . . . . . . . . Para 4-332/335, 4-338, 4-384
s 12A(4) . . . . . . . . . . . . . . . . . . . . . . . . . . . . . . . . . . . . . Para 4-336/338
s 12A(5) . . . . . . . . . . . . . . . . . . . . . . . . . . . . . . . Para 4-339/340, 4-399
s 12B . . . . . . . . . Para 4-98, 4-307, 4-312, 4-320, 4-321, 4-342, 4-345,
. . . . . . . . . . . . . . . . . . . . . . . . 4-349, 4-376, 4-389, 4-396, 4-403, 4-404
s 12B(1) . . . . . . . . . . . . . . . . . . . . . . . . . . . . . . Para 4-320, 4-341/350
s 12B(2) . . . . . . . . . . . . . . . . . . . . . . . . . . Para 4-341/350, 4-358, 4-387
s 12B(3) . . . . . . . . . . . . . . . . . . . . . . . . . . . . . . . Para 4-342, 4-351/355
s 12B(4) . . . . . . . . . . . . . . . . . . . . . . . . . . . . . . . Para 4-356/358, 4-399
s 12B(5) . . . . . . . . . . . . . . . . . . . . . . . . . . . . . . . . . . . . . Para 4-359/363
s 12B(6) . . . . . . . . . . . . . . . . . . . . . . . . . . . . . . . Para 4-361, 4-364/365
s 12B(7) . . . . . . . . . . . . . . . . . . . . . . . . . . . . . . . . . . . . . Para 4-366/367
s 12C . . . . . . . . . Para 4-98, 4-307, 4-312, 4-321, 4-345, 4-376, 4-387,
. . . . . . . . . . . . . . . . . . . . . . . . . . . . . . . . . . . 4-389, 4-396, 4-403, 4-404,
s 12C(1) . . . . . . . . . . . . . . . . . . . . . . . . . . . . . . . . . . . . . Para 4-368/378
s 12C(2) . . . . . . . . . . . . . . . . . . . . . . . . . . . . . . . Para 4-368/378, 4-387
s 12C(3) . . . . . . . . . . . . . . . . . . . . . . . . . . . . . . . . . . . . . Para 4-379/382
s 12C(4) . . . . . . . . . . . . . . . . . . . . . . . . . . . . . . . . . . . . . Para 4-383/384
s 12C(5) . . . . . . . . . . . . . . . . . . . . . . . . . . . . . . . . . . . . . Para 4-385/387
s 12C(6) . . . . . . . . . . . . . . . . . . . . . . . . . . . . . . . . . . . . . Para 4-385/387
s 12D(1) . . . . . . . . . . . . . . . . . . . . . . . . . . . . . . . . . . . . . Para 4-388/389

| | |
|---|---|
| s 12D(2) | Para 4-390/391 |
| s 12D(3) | Para 4-392/393 |
| s 12D(4) | Para 4-394/395 |
| s 13(1) | Para 4-397/399 |
| s 13(2) | Para 4-400/401 |
| s 14(1) | Para 4-402/408, 4-410 |
| s 14(2) | Para 4-409/412 |
| s 14(3) | Para 4-413/415, 4-417 |
| s 14(4) | Para 4-416/417 |
| s 14(5) | Para 4-416/417 |
| s 16 | Para 4-294, 4-419 |
| s 16(1) | Para 4-418/421 |
| s 16(2) | Para 4-419, 4-422/427 |
| s 16(3) | Para 4-419, 4-428/431 |
| s 16(4) | Para 4-432/3 |
| s 16(5) | Para 4-434/435 |
| s 17 | Para 4-427, 4-437 |
| s 17(1) | Para 4-436/445, 4-447, 4-454, 4-457 |
| s 17(2) | Para 4-446/447 |
| s 17(3) | Para 4-448/450, 4-452, 4-454 |
| s 17(4) | Para 4-451/452, 4-454 |
| s 17(5) | Para 4-449, 4-452, 4-453/455 |
| s 17(6) | Para 4-456/457 |
| s 17(7) | Para 4-427, 4-458/459 |
| s 18 | Para 4-47, 4-461, 4-475, 4-477 |
| s 18(1) | Para 4-460/473, 4-477, 4-488 |
| s 18(2) | Para 4-474/475 |
| s 18(3) | Para 4-476/477 |
| s 18(4) | Para 4-478-479 |
| s 18A | Para 3-18, 4-98, 4-99, 4-308, 4-321, 4-345, 4-480/481 |
| s 18A(1) | |
| s 18A(2) | Para 4-99 |
| s 18A(3) | Para 4-99, 4-308 |
| s 19(1) | Para 4-482/484 |
| s 19(2) | Para 4-449, 4-485/486 |
| s 19(3) | Para 4-487 |
| s 20(1) | Para 4-29, 4-37, 4-47, 4-62, 4-129, 4-488 |
| s 20(2) | Para 4-133, 4-489 |
| s 20(3) | Para 4-490 |
| s 20(4) | Para 4-83, 4-491 |
| s 20(5) | Para 4-491 |

*Table of Statutes*

```
    s 52 .......................................... Para 4-450
    s 5391) ........................................ para 4-491
    s 53(2) .................................. Para 4-27, 4-491
    s 54(1) .................... Para 4-82, 4-83, 4-133, 4-137, 4-307
    s 54(2)............................... Para 4-133, 4-137, 4-307
    s 54(3)............................... Para 4-83, 4-133, 4-137
    s 55 ........................................... Para 4-05
    s 56 ........................................... Para 4-29
    s 56(1)......................................... Para 4-29
    s 56(3)......................................... Para 4-29
    s 56(4)......................................... Para 4-29
    s 58(1)......................................... Para 3-08
    s 58(2) .................................. Para 3-08, 4-29
    s 58(3) ........................................ Para 4-29
    s 59(1)............................... Para 4-29, 4-36, 4-39
    s 59(2) .................................. Para 4-29, 4-36
    s 60(1) .... Para 3-06, 4-19, 4-23, 4-32, 4-39, 4-47, 4-50, 4-60, 4-450
    s 62(4).......................................... Para 4-05
    Part I, Sched.1 ............................ Para 4-361, 4-363
Law of Property Act 1925
    s 196......................................... Para 4-82, 4-92
Law of Property (Miscellaneous Provisions) Act 1989
    s 2 ............................................. Para 4-97
Leasehold Reform Act 1967 ........... Para 2-02, 2-08/18, 3-06, 4-18,
     ........................................... 4-65, 6-17/18
Leasehold Reform, Housing and Urban Development
    Act 1993 ............. Para 2-02, 2-11, 2-18/20, 3-11, 4-23, 4-47,
     ........................................ 5-01/5-60, 6-13/21
    s 3 ............................................. Para 5-09
    s 4 ............................................. Para 5-09
    s 7 ............................................. Para 5-16
    s 11 ............................................ Para 5-25
    s 13 ............................................ Para 5-26
    s 26 ............................................ Para 5-25
    s 27 ............................................ Para 5-25
    s 99 ............................................ para 5-29
    s 101 ........................................... Para 5-18
    s 101(1)........................................ Para 5-14
    Sched. 1 ....................................... Para 5-25
    Sched. 6, Part II ...................... Para 5-56, 5-57, 5-58
    Sched. 7 ....................................... Para 5-51
```

Sched. 9, Part III ............................... Para 5-41
Sched. 9, Part IV ............................... Para 5-42
Local Government Act 1985
    s 10 ........................................ Para 4-29
    Part IV ..................................... Para 4-29
Local Government and Housing Act 1989
    Sched. 10 .................................. Para 5-16
Local Government, Planning and Land Act 1980,
    Part XVI .................................... Para 4-29

New Towns Act 1981 ............................... Para 4-29

Police Act 1996 ................................... Para 4-29
Proceeds of Crime Act 2002 .................. Para 2-06, 4-288

Recorded Delivery Service Act 1962
    s 1, Sch .................................... Para 4-82
Rent Act 1977 ............................... Para 4-32, 4-39
Rent (Agriculture) Act 1976 .................. Para 4-32, 4-39

Trustee Act 1925
    s 34(2) .................................... Para 4-130

# Table of Statutory Instruments

Leasehold Reform (Collective Enfranchisement and
    Lease Renewal) Regulations 1993, SI 1993/2407...... Para 5-05
    Sched. 1 ........................................ Para 5-51
Leasehold Reform (Collective Enfranchisement)
    (Counter-Notices) (England) Regulations 2002,
    SI 2002/3208 .................................... Para 5-05
Right to Manage (Prescribed Particulars and Forms)
    (England) Regulations 2003, SI 2003/1988 ...... Para 5-63, 5-67
RTM Companies (Memorandum and Articles of Association
    (England) Regulations 2003, SI 2003/2120........... Para 5-66

# Introduction to Mixed Use Buildings

**1-01**
Mixed use, in the sense of a single building used for both residential and commercial purposes, is by no means a new concept. Indeed, since medieval times, craftsmen and workers have lived above their workshop, and any house built in the centre of a town before the 17th century would originally have comprised a shop or workshop below and living space above. Only in the 20th century did the space above the high street come to be left unoccupied, often due to insurance reasons.

**1-02**
But this is not the meaning of mixed use to which our book relates. For this purpose, mixed use involves the following ingredients:

- A single building.
- Owned by a single freeholder.
- Let to a number of different tenants.
- Some tenants using their premises for commercial purposes.
- Other tenants using their premises for residential purposes.

**1-03**
This kind of mixed use is a relatively recent development, and began in the 20th century, mainly post-Second World War. It originated from two separate places: first, the realisation of the Government that town centres became dark, empty places after the shops and other businesses closed, and the desire to re-introduce residential use into these areas, as supported by planning legislation, and second, the realisation by many

large retailers that the surplus space above their retail shops could be utilised for flats, thus releasing value. Retailers such as Waitrose, IKEA, Tesco, Marks & Spencer and, more recently, McDonalds, have led the development of flats above their stores since the late 1970s/early 1980s, and it is now common to see a new town-centre development with shops on the ground floor, offices above and flats above the offices.

### 1-04
It often comes as a surprise to such retailers or commercial developers of such mixed use developments that the rights of residential tenants can impinge on their ability to own, manage and dispose of the building. Those more accustomed to dealing with commercial tenants will be entirely familiar with the security of tenure and rights to a new tenancy conferred on such tenants by statute. Many residential tenants enjoy, however, in addition to security of tenure and rights to a new tenancy, far more extensive rights, enabling them compulsorily to acquire the freehold of the building as well as rights to take over the management of the residential part of the building and rights of first refusal in case of a proposed sale of the building.

### 1-05
Throughout the 20th century, legislation to protect tenants—particularly residential tenants—was introduced, but it was not until 1967 that legislation was first passed entitling long leaseholders of houses (subject to satisfying the prescribed conditions) to acquire the freeholds of their buildings compulsorily. This right was extended to long leaseholders of flats in 1993, and most of the prescribed conditions have since been removed.

### 1-06
Alongside these rights, the Landlord and Tenant Act 1987 gave residential tenants rights of first refusal where their landlord proposes to dispose of their building or an interest in it, as well as the right to apply to the court for a third party to manage the building in place of a bad landlord manager. The rights of first refusal were substantially extended in 1996, and an additional right to manage in favour of the tenants was introduced in 2002, which applied to all landlords (not just bad managers). Legislation also gives residential tenants considerable protection in relation to service charges, but this is outside the scope of our book.

# Outline of the Impact of Residential Tenants' Rights on Mixed Use Properties

**2-01**
A mixed use property, which contains both commercial and residential premises, may be affected by specialist residential legislation that can deprive a property owner of its property, the right to manage its property or the right to dispose of its property to the person and at the time it desires. Commercial tenants who occupy a mixed use property may find that their professional, institutional landlord has been replaced by a group of residential tenants, whose main concern is to minimise expenditure on the property, or that control of the management of the property is now split between the landlord and the residential tenants.

**2-02**
In this chapter, we look at the effects of the Landlord and Tenant Act 1987 (tenants' rights of first refusal), the Leasehold Reform, Housing and Urban Development Act 1993 (collective enfranchisement and the individual right to a lease extension: flats) and the Commonhold and

Leasehold Reform Act 2002 (the right to manage) in relation to mixed use buildings. We also look very briefly at the effects of the Leasehold Reform Act 1967, which deals with the right to enfranchise enjoyed by a tenant of a "house", which may include a mixture of both residential and commercial premises, but which is outside the scope of this book.

## A. The Landlord and Tenant Act 1987, Part I (as amended)—tenants' rights of first refusal

**2-03**
The Landlord and Tenant Act 1987 was originally enacted in a hurry before the 1987 General Election following the "Report of the Committee of Inquiry on the Management of Privately Owned Blocks of Flats", chaired by E.G. Nugee QC, in October 1985. It was a logical extension to the Government's policy since the Second World War to protect and benefit residential tenants as against landlords, which had included legislation providing security of tenure and restrictions on rent increases. This Act was intended to give residential tenants a right of first refusal in cases where their landlord proposed to sell the block of flats in which they lived.

**2-04**
However, the Act was extremely badly drafted, and led to a series of cases involving cynical landlords who ignored the legislation and generally got away with it. A groundswell of adverse press publicity culminated in a campaign orchestrated by the *Evening Standard* and the Leasehold Enfranchisement Association at the time of the sale of the Kensington Estate by Henry Smith's Charity and the two Court of Appeal decisions relating to it in which we were involved. This led directly to the Housing Act 1996, which substituted the majority of the sections of Part I of the 1987 Act with new sections. Unfortunately, however, the Act, as amended, remains badly drafted and unworkable in a commercial context. The two main changes wrought by the 1996 Act are that failure to comply with the Act has become a criminal offence, and an exchange of contracts is now caught by the Act.

## 2-05

The Act (as amended) prohibits a landlord from making a "relevant disposal" without first offering to make that disposal in favour of the "qualifying" residential tenants. Whilst this may not seem a problem in itself for the landlord, the need to wait for a minimum of two months after serving the "offer notices", even if the residential tenants have no intention of exercising their rights, impacts adversely on a landlord's freedom to sell the property at a particular time (perhaps before the end of a tax year, or before the end of a landlord company's accounting year) and may, in a falling property market, effectively prevent the landlord from selling the property at all.

## 2-06

Further, if the landlord ignores the residential tenants' rights and fails to serve the "offer notices" before selling the property, the landlord commits a criminal offence. The current fine is a maximum of £5,000. Where the landlord is a company, any officer who consents to or connives at the offence, or causes the same by neglect, is separately guilty of the same offence. Furthermore, the landlord's solicitors may be guilty of an offence under the Proceeds of Crime Act 2002.

## 2-07

The landlord's guilt has no effect whatsoever on the validity of the landlord's disposal. However, where the landlord fails to comply with the requirements of the 1987 Act in making a disposal, the residential tenants will have separate rights against the purchaser, and may be able to force the purchaser (or even a successor in title of the purchaser) to sell the property to them on the same terms as those on which the landlord disposed of it. Worse still, although the residential tenants' rights to acquire the property are limited in time, time does not start running against the tenants until they have been informed both that the disposal has taken place and of their rights in relation to it.

## B. The Leasehold Reform Act 1967 (as amended) — tenant's enfranchisement rights (houses)

**2-08**
Largely in order to protect homeowners from owning a wasting asset in the form of a leasehold house, the Leasehold Reform Act 1967 was enacted, enabling some leaseholders of some houses to purchase the freehold of their house (enfranchise), or to be granted a new lease expiring 50 years after the expiry of the current lease, in each case whether or not the landlord wishes this to happen.

**2-09**
It is virtually impossible for landlords to prevent qualifying tenants from exercising their rights, although landlords will receive monetary compensation for the loss of their property or a premium on the grant of an extended lease, in each case calculated in accordance with the 1967 Act.

**2-10**
Although an obvious act of expropriation, and ostensibly contrary to the European Convention on Human Rights, Protocol 1, article 1, the European Court has held (in the case of *James* v *United Kingdom* (1968) 8 ECHR 123) that compulsory enfranchisement is in the public interest in removing a social injustice.

**2-11**
The 1967 Act rights were subsequently extended by the Housing Act 1974, the Leasehold Reform, Housing and Urban Development Act 1993 and the Housing Act 1996 to include larger and different groups of houses. In 2002, the Commonhold and Leasehold Reform Act 2002 abolished (for most purposes) the residence requirement and the rateable value/low rent tests (except for valuation purposes), thus making it much easier for leaseholders to exercise their rights. As a result, the rights have been extended beyond their initial purpose — to allow homeowners to purchase their own homes — enabling investor leaseholders and even companies to qualify.

## 2-12
The definition of a house under the 1967 Act is "any building designed or adapted for living in and reasonably so called". The definition has been held to include a shop with residential premises above where the whole building is let on a single lease — *Tandon* v *Trustees of Spurgeons Homes* [1982] 2 EGLR 73.

## 2-13
In deciding whether or not the building is "designed or adapted for living in and reasonably so called" the key question is "first to consider the property as it was initially built: for what purpose was it originally designed? That is the natural meaning of the word 'designed', which is a past participle" — per Lord Neuberger, *Boss Holdings Ltd* v *Grosvenor West End Properties* [2008] UKHL 5; [2008] 05 EG 167 (CS).

## 2-14
In the *Boss* case, the building was originally built in the 18th century as a house but subsequently came to be used partially for commercial purposes. At the time of the service of the notice to enfranchise, the building was unoccupied and, indeed, incapable of being occupied because it was so dilapidated. Lord Neuberger, giving the judgment of the Court, held that the building was a house within the meaning of the 1967 Act, because it was designed for living in when first built and nothing that had happened since had changed this.

## 2-15
Lord Neuberger also made obiter comments relating to whether a building would be a "house" if it had originally been designed for living in but had subsequently been adapted to another use. He said:

> As a matter of literal language, such a property would be a house, because 'designed' and 'adapted' appear to be alternative qualifying requirements...

## 2-16
This raised the spectre that a building originally designed as a house but now used entirely (or mainly) for commercial purposes may fall within the 1967 Act. However, in the subsequent Court of Appeal case

of *Grosvenor Estates Ltd* v *Prospect Estates Ltd* [2008] EWCA Civ 1281, which related to a building originally designed as a dwelling but subsequently occupied under the terms of a predominantly (88.5%) commercial lease, the Court of Appeal, making no reference to the judgment of the House of Lords in *Boss*, held that such a building could not be a "house" within the meaning of the 1967 Act. Their decision was on the basis that a building primarily used for commercial purposes could not be a house "reasonably so called".

**2-17**
If *Prospect* goes to the House of Lords on appeal, the position should be clarified once and for all in respect of mixed use buildings, but until that happens, anyone seeking to purchase the freehold of a building originally designed as a dwelling should be aware that a long leaseholder may have enfranchisement rights under the 1967 Act even if the building is currently used wholly or mainly for commercial purposes.

## C. The Leasehold Reform, Housing and Urban Development Act 1993 (as amended) — collective enfranchisement (flats)

**2-18**
The ability of leaseholders of houses to enfranchise under the Leasehold Reform Act 1967 was extended to flats, thus enabling flat leaseholders also to protect what was otherwise considered to be a depreciating asset. The Leasehold Reform, Housing & Urban Development Act 1993, as amended by the Commonhold and Leasehold Reform Act 2002, provides for a sufficient number of qualifying leaseholders, acting together, to acquire the freehold of the property of which their flats form part (even where this is a mixed use property), whether the landlord wishes this or not. Further, individual tenants are entitled to require the grant of a new lease of their individual flat. The landlord is compensated in each case on the basis set out in the legislation.

**2-19**
The 1993 Act cannot, for most practical purposes, be excluded in part or whole. So, for example, a lease within the 1993 Act cannot require

tenants to refrain from exercising their rights (if any) under that Act. Similarly, it cannot require tenants to offer to surrender their lease, nor impose any kind of penalty, in the event that the tenants seek to exercise such rights. The 1993 Act also prohibits the increase in the value of the freehold by, for example, the creation of an overriding lease.

**2-20**
If there are superior leaseholders of the building between the qualifying leaseholders and the freeholder, their interests will effectively be "hoovered up" as part of the collective enfranchisement, and the purchaser nominated by the qualifying leaseholders will become their direct landlord.

## D. The Commonhold and Leasehold Reform Act 2002 — the right to manage

**2-21**
Since 1987, by virtue of the Landlord and Tenant Act 1987, Part II, leaseholders have been able to apply to the court for the appointment of a manager, and even to acquire the landlord's interest, even against the landlord's wishes, if they could establish the required landlord default.

**2-22**
The Commonhold and Leasehold Reform Act 2002, in contrast, allows residential leaseholders to take over the management of their block of flats (or the residential part of their mixed use building) without having to show default on the part of either the landlord or any management company. The 2002 Act takes control from the landlord, and places most of the landlord's duties and liabilities on the residential leaseholders in the form of the Right to Manage Company. The landlord thus loses control over the standard of management, repairs, major works, and improvements, provision of services, budgets and reserve funds for the building, without any right to compensation.

**2-23**
In a purely residential building, this may have an adverse impact on the value of the landlord's reversionary interest, with the possibility for conflict between the landlord's essentially long term interest and

the residential leaseholders' sometimes shorter term focus. However, in a mixed use building, the effect on the landlord's interest may be greater. Even though the rights of the residential leaseholders do not extend to the commercial parts of the building, control of such parts remaining with the landlord, the legislation does not deal with the determination of disputes between the landlord/commercial tenants and the residential leaseholders with regard to such matters as the overall maintenance and appearance of the building as a whole, maintenance of shared common parts and services and the remedying of inherent defects that affect the building as a whole.

# The Landlord and Tenant Act 1987, Part I (as amended) —Tenants' Rights of First Refusal

## A. Determining whether or not the 1987 Act applies

**3-01**
Section 1(1) of the Landlord and Tenant Act 1987 Act prohibits a landlord from making a "relevant disposal" affecting any premises to which at the time of the disposal the Act applies unless the landlord has previously served notice pursuant to section 5 of the Act on the qualifying tenants of those premises, and unless the disposal is made in accordance with the requirements of sections 6 to 10 of the Act.

**3-02**
In order to decide whether or not a proposed disposal is caught by the 1987 Act, it is necessary to answer four questions, namely:

1. Are the premises to be disposed of affected by the Act?
2. Is the landlord exempt from needing to comply with the Act?
3. Are there sufficient "qualifying tenants" in the premises?
4. Is the proposed transaction a "relevant disposal"?

**3-03**
Provided that at least one of the above questions can be answered in the negative, the landlord is not obliged to comply with the 1987 Act in carrying out the proposed transaction. The questions can be answered in any order, but it makes sense to consider first any question that is likely to take the proposed disposal outside of the Act. For example, if it is believed that the landlord enjoys an exemption from the Act, then it makes sense to deal with the second question first; if the landlord proves to be exempt, then the other questions do not need to be answered.

## *The premises*

**3-04**
The 1987 Act applies to premises if all three of the following apply:

- they consist of the whole or part of a building
- they contain two or more flats held by qualifying tenants
- the number of such flats exceeds 50% of the total number of flats in the building.

**3-05**
In the case of a mixed use building, the 1987 Act will only apply if the internal floor area of the residential part represents at least 50% of the internal floor area of the premises as a whole. A building that contains 50.01% commercial will therefore fall outside of the Act, whereas a building that contains 50% commercial will fall within the Act. Common parts are ignored for the purposes of this calculation. Where the commercial part of a building appears to represent approximately half of the space within the building, it may be worth the landlord's while having a detailed professional survey of the building carried out, as the extent of the common parts within the residential part of the building may be enough to tip the balance and take the building outside of the Act.

**3-06**
"Flat" is defined by section 60(1) of the 1987 Act as a separate set of premises (whether or not on the same floor) which forms part of a building, is divided horizontally from some other part of that building

and is constructed or adapted for the purposes of a dwelling. A building that contains a single flat will never fall within the 1987 Act, but may fall within the Leasehold Reform Act 1967 (as amended).

**3-07**
"Building" is not defined. Two particularly difficult areas to deal with, in the context of the 1987 Act, are terraces (where the issue is whether it is the terrace as a whole that qualifies as the "building", or whether each individual property within the terrace is itself "a building") and building estates (where the issue is whether each individual block of flats is a separate "building" or whether the estate as a whole can be regarded as a single "building scheme"). See Chapter 4, para 4.17 in relation to the meaning of "building".

## The landlord

**3-08**
Every landlord is caught by the 1987 Act, except for:

- an "exempt landlord". This definition, contained in section 58(1) of the Act, provides a list of public bodies, including local and police authorities, development corporations, urban development corporations and registered housing associations

- a "resident landlord". This exception does not apply to purpose built blocks of flats, and the definition of "resident landlord" contained in section 58(2) of the Act refers to a landlord who occupies a flat in the building as the landlord's only or principal residence, and has so occupied it for at least the immediately previous 12 months

- a landlord who is not the immediate landlord of the qualifying tenants, unless the intermediate tenancy is for a term of less than seven years, or is terminable within the first seven years at the option of the superior landlord. Note that the grant of such an immediate tenancy, perhaps with the intention of avoiding the need for compliance with the Act, will itself be a "relevant disposal" and so will be caught by the Act

- the Crown.

## Qualifying tenants

### 3-09
Every tenant is a qualifying tenant for the purposes of the 1987 Act, including companies, persons who do not reside in the relevant flat and Rent Act tenants, except for the following:

- a tenant under a protected shorthold tenancy

- a tenant under a business tenancy to which the Landlord and Tenant Act 1954, Part II applies

- a tenant under a tenancy terminable on the ending of the tenant's employment

- a tenant under an assured tenancy (including an assured shorthold tenancy) or under an assured agricultural occupancy

- a tenant who is the tenant (under one or more leases) of three or more flats in the building (excluding any flat that falls within one of the above four exceptions)

- a tenant whose landlord is also a qualifying tenant of that flat.

### 3-10
More than half of the flats in the building must be let to qualifying tenants, which means that in the case of a building containing only two flats, both must be let to qualifying tenants.

## Relevant disposal

### 3-11
Any disposal by the landlord of any legal or equitable estate or interest (including the surrender of a tenancy) in any premises to which the 1987 Act applies will be caught by the Act, apart from the following exceptions:

- The grant of a tenancy of a single flat (including any garden, outhouse or similar appurtenance enjoyed with the flat, not being a common part of the building). Note that the grant of a tenancy of a single commercial unit is not specifically exempted from the 1987 Act, although it is not clear how the requirements of the 1987 Act would interact with the requirements of the Landlord and Tenant Act 1954, Part II.

- The disposal of any interest of a beneficiary in settled land.

- The disposal of any easement, rent charge or other form of incorporeal hereditament.

- The creation of a mortgage or charge.

- A disposal to a trustee in bankruptcy or the liquidator of a company.

- Various disposals pursuant to a court order made in specified matrimonial, family or inheritance proceedings.

- A disposal pursuant to a compulsory purchase order or an agreement entered into by way of compromise to avoid a compulsory purchase order being made or carried into effect.

- A disposal by way of collective enfranchisement or lease renewal under the Leasehold Reform, Housing and Urban Development Act 1993.

- A disposal by way of gift to a member of the landlord's family or to a charity.

- A disposal of functional land by one charity to a second charity which intends to use that land as functional land.

- A disposal of land held on trust in connection with the appointment or discharge of a trustee.

- A disposal by two or more members of a family to fewer of their number, or to a different combination of family members (to include at least one of the transferors).

- A disposal in pursuance of a contract, option or right of pre-emption binding on the landlord (except as provided by section 8D of the 1987 Act). Note that the exchange of contracts or grant of the option or right of pre-emption will, however, be a relevant disposal.

- The surrender of a tenancy pursuant to a covenant, condition or agreement contained in that tenancy.

- A disposal to the Crown.

- A disposal between associated companies which have been associated for at least two years.

- A disposal under the terms of a will or the law relating to intestacy.

## B. If the 1987 Act does apply to the intended disposal, what must the landlord do before making the disposal?

### Section 5 requirements

**3-12**

Under section 5 of the 1987 Act, the landlord must serve an offer notice on the qualifying tenants of the flats contained in the premises. The offer notice will be an offer made "subject to contract", and each offer notice must comply with the requirements set out in that subsection of section 5 relating to the particular proposed disposal:

- Section 5A: a proposed contract to create or transfer an estate or interest in land (not applicable to the grant of an option or right of pre-emption, which is covered by section 5C).

- Section 5B: a proposed sale at a public auction to be held in England and Wales.

- Section 5C: a proposed grant of an option or right of pre-emption.

- Section 5D: a proposed disposal not made in pursuance of a contract, option or right of pre-emption.

- Section 5E: a proposed disposal under sections 5A, 5B, 5C or 5D where the disposal is to be (wholly or partly) for a non-monetary consideration.

## 3-13
Where the proposed disposal relates to more than one building, the landlord is obliged to sever the transaction so as to deal with each building separately. This is particularly difficult, given the absence of a definition of "building", as previously stated.

## 3-14
Where section 5A applies, the notice must:

- contain particulars of the principal terms of the proposed disposal, including in particular the property and the estate or interest therein proposed to be disposed of and the principal terms of the intended contract (including deposit and consideration required)

- state that the notice constitutes an offer by the landlord to enter into a contract on the above terms, which may be accepted by the requisite majority of qualifying tenants

- specify a period within which the offer may be accepted, being not less than two months, starting with the date of service of the notice

- specify a further period of not less than two months, within which the tenants may nominate a person or persons to purchase the property under section 6.

## 3-15
Similar requirements apply to sections 5B–5E, but in the case of section 5B (a sale by public auction in England and Wales), the timescale for service of the notice is not less than four and not more than six months before the date of the auction, and the proposed price does not need to be specified (although if the notice does not state the time and place of the auction and the name of the auctioneers, then the landlord must, at least 28 days before the auction date, serve a further notice on the requisite majority of the qualifying tenants, giving those particulars). Further, the tenants' rights are to step into the shoes of the successful bidder at the auction; they have no separate right to acquire before or at the auction, and if the property is not sold at the auction, their rights will come to nothing.

**3-16**
Once the landlord has served an offer notice under section 5, the landlord must not make the proposed disposal other than to the tenants' nominated purchaser during the period (being a minimum of two months) for acceptance specified in the offer notice (or such longer period as may be agreed between the landlord and the requisite majority of the qualifying tenants). A landlord may be tempted to ignore this requirement, particularly where a negative response has been received from all (or the vast majority of) the tenants at an early stage in the two-month period. However, in addition to being a criminal offence, the landlord needs to be aware that the tenants can change their responses as many times as they choose during the two-month period.

## C. What happens then?

**3-17**
If the tenants wish to exercise their rights of first refusal, they must serve an acceptance notice on the landlord and nominate a person to take the purchase on their behalf. In the case of a sale at public auction, there is an additional requirement for the tenants to comply with, namely the nominated purchaser must serve notice on the landlord at least 28 days before the date of the auction, confirming that the norminated purchaser wishes to be able to exercise the right to step into the shoes of the successful bidder at the auction.

**3-18**
Section 18A of the 1987 Act, inserted by the Housing Act 1996, defines the "requisite majority of qualifying tenants of the constituent flats", in the context of a response to a section 5 offer notice, to mean qualifying tenants of constituent flats with more than 50% of the available votes. In other words, there must be a positive response from more than 50% of the total number of those flats in the building which are let to qualifying tenants on the date when the acceptance period expires, if the tenants are to exercise their rights under the 1987 Act. There is one available vote for each such flat, although it is not clear what would happen if two people were jointly the tenant of a flat, and their votes were different.

## 3-19
If no acceptance notice is served by the requisite majority of the qualifying tenants within the acceptance period specified in the section 5 notices, then the landlord is free to proceed with the proposed disposal (so long as the disposal is of the interest specified in the offer notices) to any third party during the 12-month period commencing at the end of the acceptance period, but subject to the following:

- Where section 5B applied (sale by auction), the disposal must be made by the landlord at a public auction and the other terms of the disposal must correspond to those specified in the section 5 notices.

- In all other cases, the deposit and consideration must be not less than, and the other terms must correspond to, those specified in the section 5 notices.

## 3-20
If any acceptance notice is served by the requisite majority of the qualifying tenants within the acceptance period specified in the section 5 notices, then the landlord must not proceed with the proposed disposal, other than to the person nominated by the requisite majority of the qualifying tenants, until the expiry of the period specified in the section 5 notices for nomination (or such later date as the landlord and the tenants may agree).

# D. After acceptance and nomination

## 3-21
Following service of an acceptance notice by the requisite majority of the qualifying tenants and nomination, the landlord may choose one of only two alternatives and must implement this choice within one month of the date of service of notice of nomination. The landlord's choices are either to serve notice on the nominated person of the intention not to proceed with the disposal, or to proceed with the sale to the nominated person.

## 3-22
The landlord may initially choose to proceed but subsequently choose to serve a notice of withdrawal (although there may be adverse cost

consequences to the landlord if this happens). A further possibility is withdrawal by the nominated person. Finally, if the premises cease to be premises to which the 1987 Act applies, the landlord may serve notice to that effect on the qualifying tenants, whereupon the Act will cease to apply to the landlord's proposed disposal and the landlord will be free to proceed with it.

## Withdrawal by the landlord

**3-23**
As the tenants' rights under the 1987 Act are rights of pre-emption arising from the landlord's proposal to dispose, rather than a right to buy at the tenants' option (as in the case of the right to enfranchise), the landlord cannot be compelled to proceed with the disposal, even if the requisite majority of the qualifying tenants have taken all necessary steps to exercise their rights under the Act. Once the tenants have served their acceptance and nomination notices, the landlord cannot proceed with the disposal except to the nominated person (unless the nominated person withdraws or the landlord's offer lapses and the landlord serves notice to that effect), but the landlord cannot be compelled to make the disposal at all: see Chapter 4, para 4-178.

**3-24**
If the landlord serves notice of withdrawal, the landlord must not dispose of the estate or interest in the premises that was originally proposed to be disposed of (whether on the originally proposed terms or on any other terms) for a period of 12 months, beginning with the date of service of the notice of withdrawal. However, the landlord may dispose of any other estate or interest in the premises during that period, subject to first complying with the 1987 Act.

## Proceeding with the disposal

**3-25**
Where the landlord chooses to proceed with the sale to the nominated person, the provisions of section 8A of the 1987 Act will apply, except where the proposed disposal was to be by way of sale at a public auction in England and Wales, in which case the provisions of section 8B apply.

## Section 8A — disposals (not at public auction)
### 3-26
Within one month of the date of service of the nomination notice, the landlord shall send a draft contract to the nominated person, providing for the proposed disposal on the terms set out in the offer notice. Failure to comply with this obligation means that the landlord is deemed to have served notice of withdrawal at the end of the one-month period.

### 3-27
Within two months from the date on which the draft contract is sent to the nominated person (or such longer period as may be agreed with the landlord), the nominated person shall either serve notice on the landlord stating that there is no longer an intention to proceed, or sign the contract and send it to the landlord together with the deposit required by the contract (or 10% of the consideration, whichever is the smaller amount). If the nominated person does neither (or serves notice that there is no longer an intention to proceed) then the norminated person is deemed to have served notice of withdrawal.

### 3-28
If the landlord receives a contract and any required deposit from the nominated person but fails to exchange contracts within seven days thereafter, then the landlord is deemed to have served notice of withdrawal at the end of that period.

## Section 8B — disposals at public auction
### 3-29
Provided that the nominated person has notified the landlord not less than 28 days before the date of the auction that the norminated person wishes to be able to step into the shoes of the successful bidder at the auction, the landlord must supply a copy of any contract for sale entered into at the auction to the nominated person within seven days of the auction. The nominated person then has 28 days from the date on which the copy contract was sent to serve notice on the landlord, accepting the terms of the contract, and to fulfil any conditions required to be fulfilled by the purchaser on exchange. If the nominated person does both, the nominated person will step into the shoes of the successful bidder at the auction.

**3-30**
If the nominated person fails within the second 28-day period to accept the terms of the contract and to fulfil the conditions therein, then the nominated person is deemed to have served notice of withdrawal at the end of that 28-day period.

## Withdrawal by the nominated person

**3-31**
The nominated person is free to serve notice of withdrawal on the landlord at any time when the landlord is obliged to proceed, indicating the intention no longer to proceed with the proposed acquisition, and must serve notice of withdrawal immediately when becoming aware that the number of qualifying tenants wanting to proceed with the acquisition is less than the requisite majority of qualifying tenants. This requirement does not apply once contracts have been exchanged between the landlord and the nominated person.

**3-32**
After service of a notice of withdrawal, the landlord is free to make the proposed disposal to any third party during the 12-month period beginning with the date of service of the notice of withdrawal. However, where the proposed disposal was to be at public auction in England and Wales, the actual disposal must be by means of a sale at a public auction and otherwise on the terms set out in the offer notice. Where the proposed disposal was not to be at such a public auction, the actual disposal must be for a consideration (and have a deposit) which is not less than that set out in the offer notice (or any higher amount agreed between the landlord and the nominated person) and otherwise must be on the terms set out in the offer notice.

## Lapse of landlord's offer

**3-33**
If after the landlord has served an offer notice, the premises cease to fall within the 1987 Act, the landlord may serve notice on the qualifying tenants, stating that the premises no longer fall within the Act and that the offer notice (and anything done pursuant to it) is to be treated as not having been served or done. On service of such a notice,

the Act will cease to apply to the proposed disposal, but so long as the landlord fails to serve such a notice, the Act will continue to apply to the proposed disposal.

**3-34**
The above provisions relating to service of notice of lapse do not apply once contracts have been exchanged between the landlord and the nominated person.

# E. Exercising rights against a subsequent purchaser (where the 1987 Act has not been complied with)

**3-35**
Where the vendor has made a relevant disposal of premises falling within the 1987 Act and has failed to comply with the Act in relation to the service of section 5 notices and/or in relation to any subsequent requirements, the requisite majority of qualifying tenants have rights against the purchaser to obtain information in relation to the disposal and then to step into the purchaser's shoes under the contract, or compel the purchaser to transfer the property to their nominated purchaser, or compel the grant of a new tenancy, whichever is applicable in the circumstances.

**3-36**
The requisite majority of qualifying tenants may serve notice on the purchaser, requiring particulars of the terms on which the disposal was made (including deposit and consideration required) and the date of the disposal, which the purchaser must comply with within one month of service of that notice. The notice must be served before the end of the four-month period, beginning with the date by which notices under section 3A of the Landlord and Tenant Act 1985 relating to the disposal have been served on the requisite majority of the qualifying tenants or, where section 3A does not apply, beginning on the date on which the requisite majority of qualifying tenants are served with documents indicating that the disposal has taken place and alerting them to the existence of their rights under the 1987 Act and the time within which such rights must be exercised. The notice must also be in writing and

specify the names of the persons on whose behalf it is served and the addresses of their flats, together with the name and address of the person to whom the items requested by the notice are to be given, on behalf of the tenants.

## 3-37
The requisite majority of qualifying tenants then have a period of six months from the date on which the purchaser complies with the tenants' notice seeking information (or six months from the date when the tenants were notified of the disposal, if the tenants did not serve a notice seeking information) in which to serve a purchase notice, electing to take the benefit of the contract, or the property, or the tenancy, whichever is appropriate in the particular case.

## 3.38 Flowchart A: Must the landlord comply with the 1987 Act in making a proposed disposal?

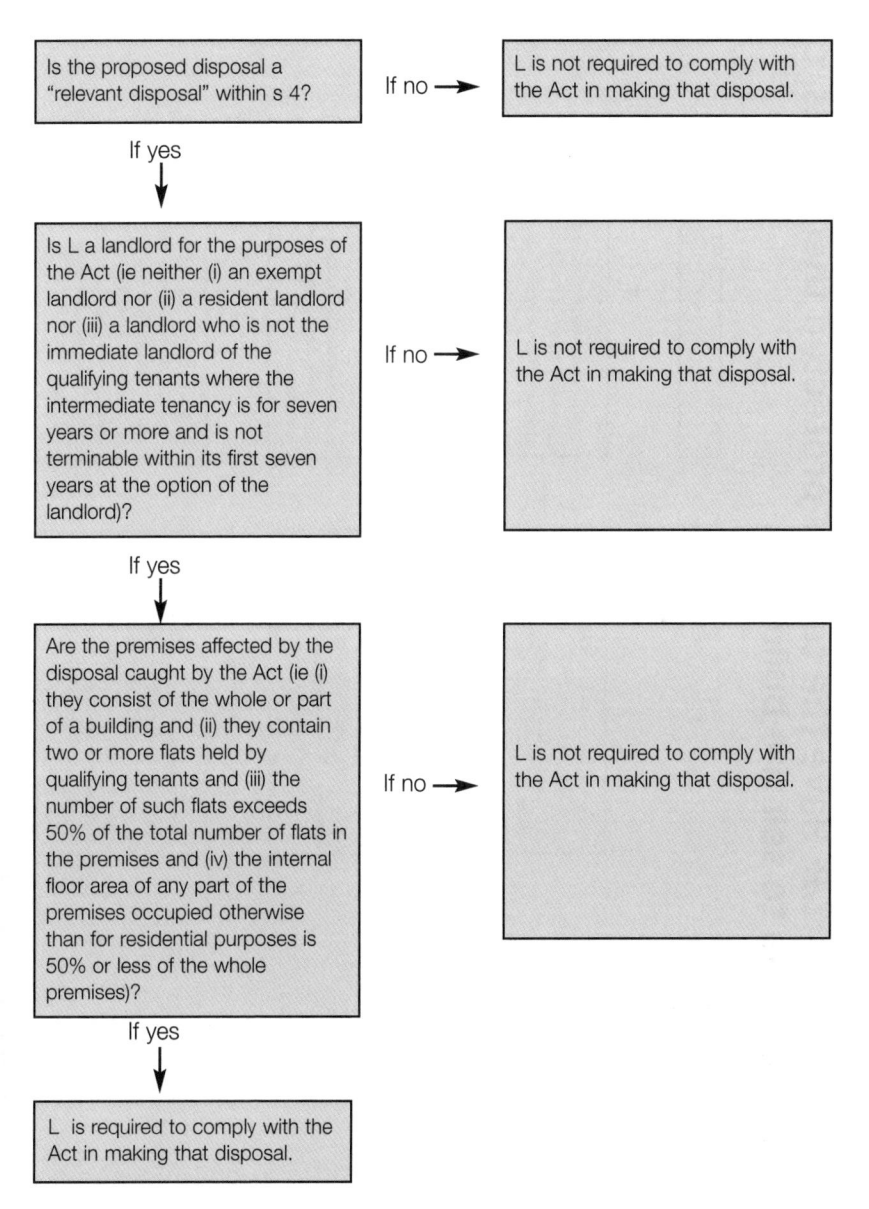

Is the proposed disposal a "relevant disposal" within s 4?

If no → L is not required to comply with the Act in making that disposal.

If yes ↓

Is L a landlord for the purposes of the Act (ie neither (i) an exempt landlord nor (ii) a resident landlord nor (iii) a landlord who is not the immediate landlord of the qualifying tenants where the intermediate tenancy is for seven years or more and is not terminable within its first seven years at the option of the landlord)?

If no → L is not required to comply with the Act in making that disposal.

If yes ↓

Are the premises affected by the disposal caught by the Act (ie (i) they consist of the whole or part of a building and (ii) they contain two or more flats held by qualifying tenants and (iii) the number of such flats exceeds 50% of the total number of flats in the premises and (iv) the internal floor area of any part of the premises occupied otherwise than for residential purposes is 50% or less of the whole premises)?

If no → L is not required to comply with the Act in making that disposal.

If yes ↓

L is required to comply with the Act in making that disposal.

## 3.39 Flowchart B: Have the tenants protected their rights, or is the landlord free to sell the building?

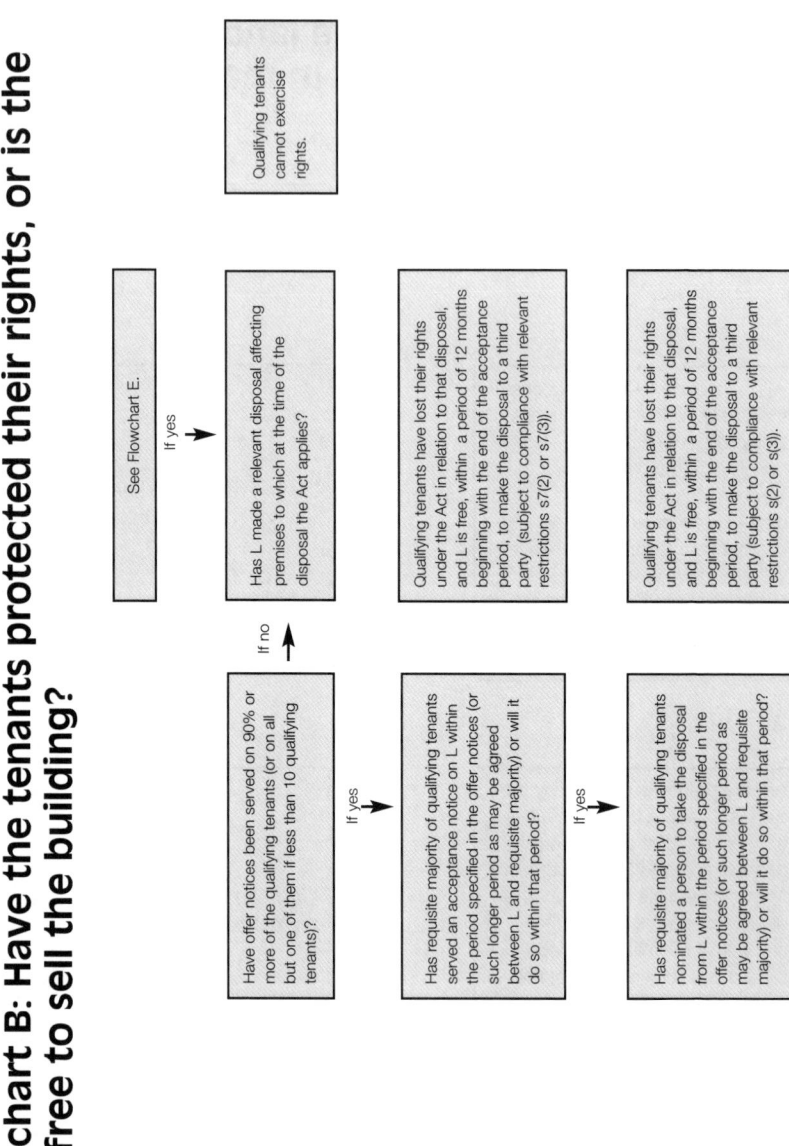

**Has L made a relevant disposal affecting premises to which at the time of the disposal the Act applies?**

If yes → See Flowchart E.

If no ↓

**Have offer notices been served on 90% or more of the qualifying tenants (or on all but one of them if less than 10 qualifying tenants)?**

If no → Qualifying tenants cannot exercise rights.

If yes ↓

**Has requisite majority of qualifying tenants served an acceptance notice on L within the period specified in the offer notices (or such longer period as may be agreed between L and requisite majority) or will it do so within that period?**

If no → Qualifying tenants have lost their rights under the Act in relation to that disposal, and L is free, within a period of 12 months beginning with the end of the acceptance period, to make the disposal to a third party (subject to compliance with relevant restrictions s7(2) or s7(3)).

If yes ↓

**Has requisite majority of qualifying tenants nominated a person to take the disposal from L within the period specified in the offer notices (or such longer period as may be agreed between L and requisite majority) or will it do so within that period?**

If no → Qualifying tenants have lost their rights under the Act in relation to that disposal, and L is free, within a period of 12 months beginning with the end of the acceptance period, to make the disposal to a third party (subject to compliance with relevant restrictions s(2) or s(3)).

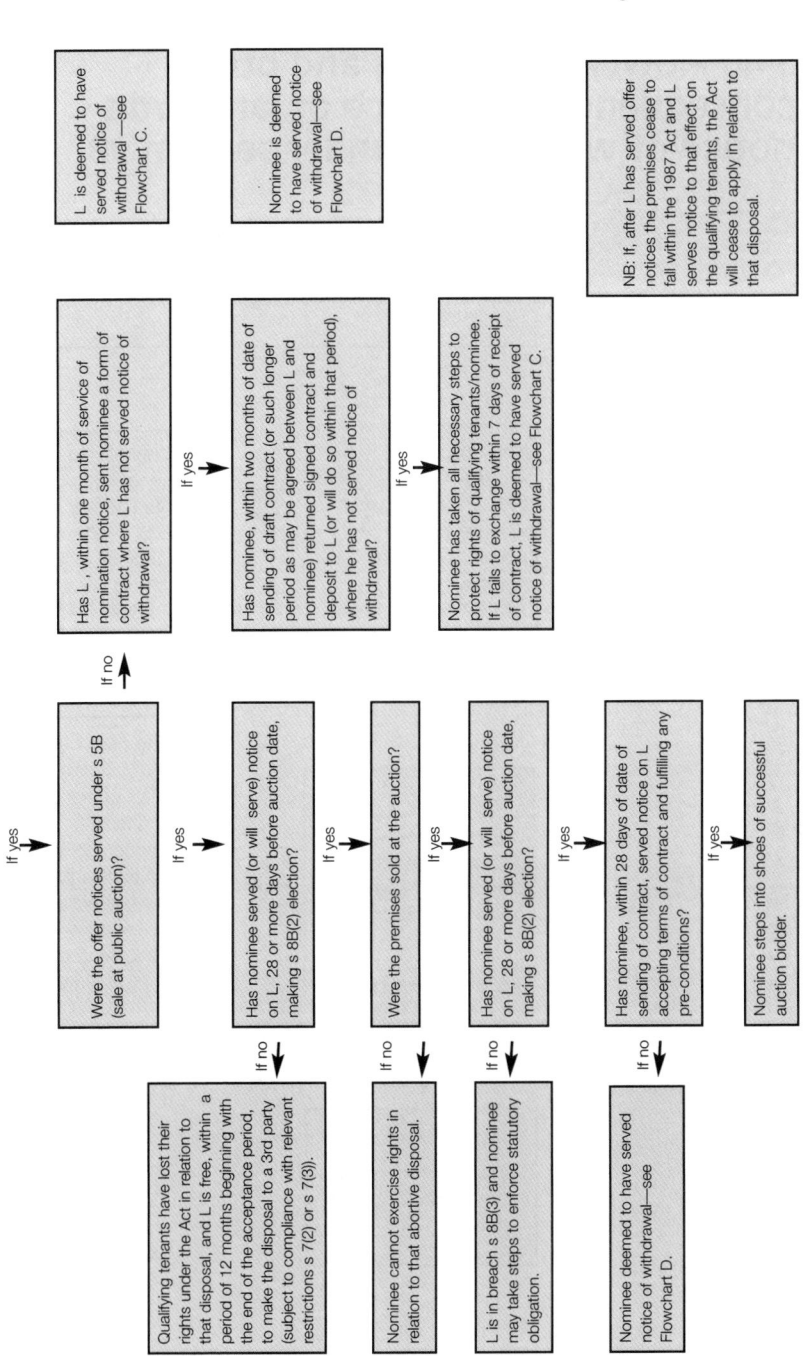

## Tenants' Rights of First Refusal

**Were the offer notices served under s 5B (sale at public auction)?**

If yes → Has L, within one month of service of nomination notice, sent nominee a form of contract where L has not served notice of withdrawal?

If yes → L is deemed to have served notice of withdrawal —see Flowchart C.

If no ↑ (back up)

If yes (down) → Has nominee, within two months of date of sending of draft contract (or such longer period as may be agreed between L and nominee) returned signed contract and deposit to L (or will do so within that period), where he has not served notice of withdrawal?

If yes → Nominee is deemed to have served notice of withdrawal —see Flowchart D.

If yes → Nominee has taken all necessary steps to protect rights of qualifying tenants/nominee. If L fails to exchange within 7 days of receipt of contract, L is deemed to have served notice of withdrawal—see Flowchart C.

**If no** (from top question) ↓

Qualifying tenants have lost their rights under the Act in relation to that disposal, and L is free, within a period of 12 months beginning with the end of the acceptance period, to make the disposal to a 3rd party (subject to compliance with relevant restrictions s 7(2) or s 7(3)).

---

Has nominee served (or will serve) notice on L, 28 or more days before auction date, making s 8B(2) election?

If yes → Were the premises sold at the auction?

If no → Nominee cannot exercise rights in relation to that abortive disposal.

If yes → Has nominee served (or will serve) notice on L, 28 or more days before auction date, making s 8B(2) election?

If no → L is in breach s 8B(3) and nominee may take steps to enforce statutory obligation.

If yes → Has nominee, within 28 days of date of sending of contract, served notice on L accepting terms of contract and fulfilling any pre-conditions?

If no → Nominee deemed to have served notice of withdrawal—see Flowchart D.

If yes → Nominee steps into shoes of successful auction bidder.

---

**NB:** If, after L has served offer notices the premises cease to fall within the 1987 Act and L serves notice to that effect on the qualifying tenants, the Act will cease to apply in relation to that disposal.

27

*Mixed Use and Residential Tenants' Rights*

## 3-40 Flowchart C: Costs and other consequences of service of landlord's notice of withdrawal under section 9B

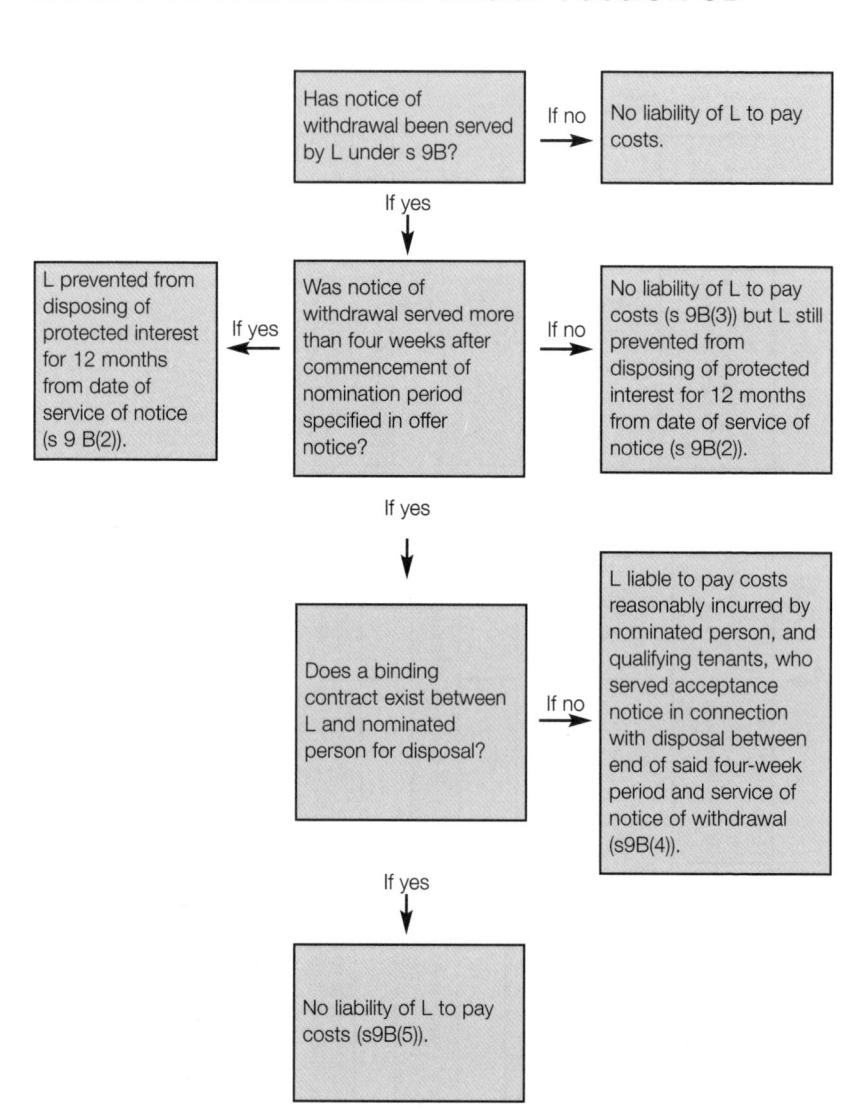

```
                    ┌─────────────────────┐
                    │ Has notice of       │  If no   ┌──────────────────┐
                    │ withdrawal been     │ ───────► │ No liability of L│
                    │ served by L under   │          │ to pay costs.    │
                    │ s 9B?               │          └──────────────────┘
                    └─────────────────────┘
                              │ If yes
                              ▼
┌──────────────────┐ ┌─────────────────────┐          ┌──────────────────────┐
│ L prevented from │ │ Was notice of       │          │ No liability of L to │
│ disposing of     │ │ withdrawal served   │          │ pay costs (s 9B(3)) │
│ protected        │◄│ more than four      │  If no   │ but L still          │
│ interest for 12  │ │ weeks after         │ ───────► │ prevented from       │
│ months from date │ │ commencement of     │          │ disposing of         │
│ of service of    │ │ nomination period   │          │ protected interest   │
│ notice (s 9 B(2))│ │ specified in offer  │          │ for 12 months from   │
│                  │ │ notice?             │          │ date of service of   │
└──────────────────┘ └─────────────────────┘          │ notice (s 9B(2)).    │
                              │ If yes                 └──────────────────────┘
                              ▼
                                                      ┌──────────────────────┐
                                                      │ L liable to pay      │
                                                      │ costs reasonably     │
                    ┌─────────────────────┐          │ incurred by          │
                    │ Does a binding      │          │ nominated person,    │
                    │ contract exist      │  If no   │ and qualifying       │
                    │ between L and       │ ───────► │ tenants, who served  │
                    │ nominated person    │          │ acceptance notice in │
                    │ for disposal?       │          │ connection with      │
                    └─────────────────────┘          │ disposal between end │
                              │ If yes                │ of said four-week    │
                              ▼                       │ period and service   │
                                                      │ of notice of         │
                    ┌─────────────────────┐          │ withdrawal (s9B(4)). │
                    │ No liability of L to│          └──────────────────────┘
                    │ pay costs (s9B(5)). │
                    └─────────────────────┘
```

# 3-41 Flowchart D: Costs and other consequences of service of nominated person's notice of withdrawal under section 9A

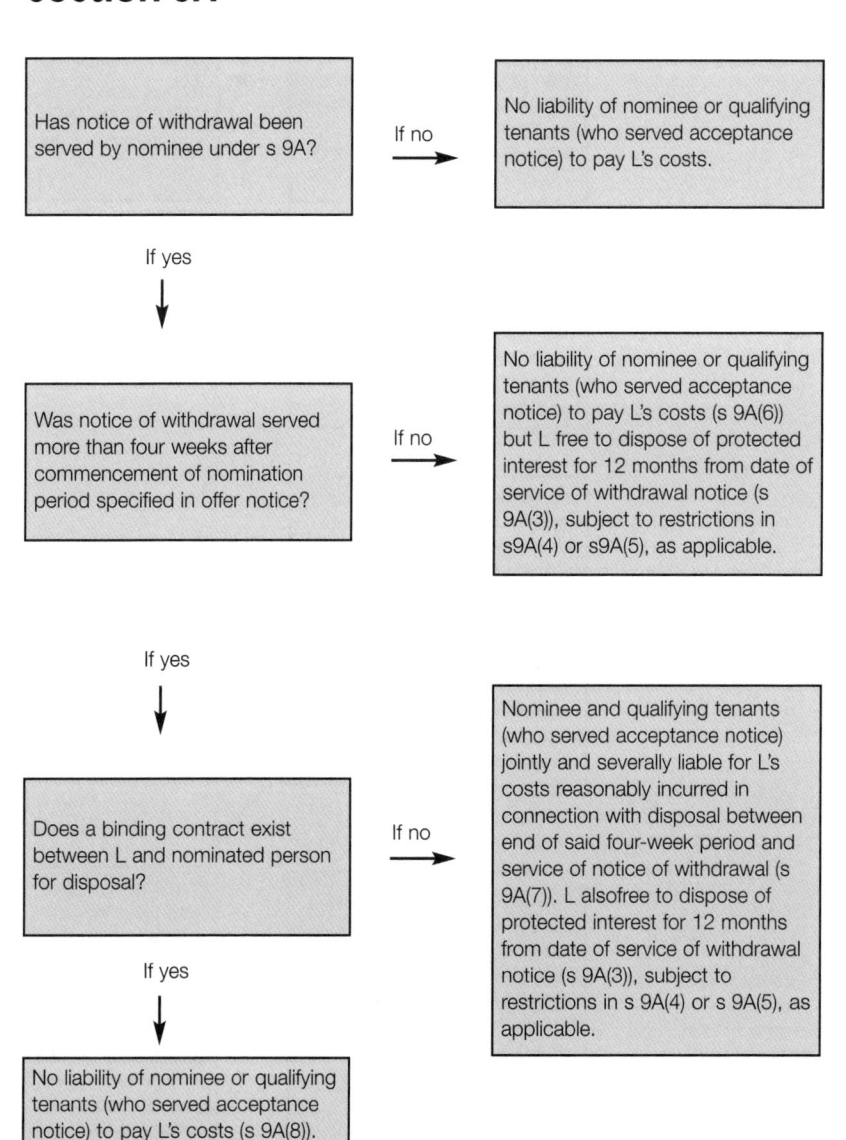

Has notice of withdrawal been served by nominee under s 9A?

If no → No liability of nominee or qualifying tenants (who served acceptance notice) to pay L's costs.

If yes ↓

Was notice of withdrawal served more than four weeks after commencement of nomination period specified in offer notice?

If no → No liability of nominee or qualifying tenants (who served acceptance notice) to pay L's costs (s 9A(6)) but L free to dispose of protected interest for 12 months from date of service of withdrawal notice (s 9A(3)), subject to restrictions in s9A(4) or s9A(5), as applicable.

If yes ↓

Does a binding contract exist between L and nominated person for disposal?

If no → Nominee and qualifying tenants (who served acceptance notice) jointly and severally liable for L's costs reasonably incurred in connection with disposal between end of said four-week period and service of notice of withdrawal (s 9A(7)). L also free to dispose of protected interest for 12 months from date of service of withdrawal notice (s 9A(3)), subject to restrictions in s 9A(4) or s 9A(5), as applicable.

If yes ↓

No liability of nominee or qualifying tenants (who served acceptance notice) to pay L's costs (s 9A(8)).

*Mixed Use and Residential Tenants' Rights*

# 3-42 Flowchart E: Tenants' rights following unauthorised disposal by landlord

```
┌─────────────────────────────────┐
│ Have more than four months      │          ┌─────────────────────────┐
│ elapsed since requisite majority│          │ The requisite majority  │
│ of qualifying tenants were      │          │ of qualifying tenants   │
│ served with (i) documents       │  If yes  │ can no longer seek      │
│ indicating that original        │ ────────►│ information about the   │
│ disposal has taken place,       │          │ disposal under s        │
│ alerting them to existence of   │          │ 11A(1), but may still   │
│ their rights and time period    │          │ execrcise rights under  │
│ for exercise, or (ii) notices   │          │ s 12—see below.         │
│ under s 3A of the Landlord      │          └─────────────────────────┘
│ & Tenant Act 1985?              │
└─────────────────────────────────┘
                │
              If no
                ▼
┌─────────────────────────────────┐
│ The requisite majority of       │
│ qualifying tenants can seek     │
│ information about the disposal  │
│ under s 11A(1), and may         │
│ execrcise rights under s 12—    │
│ see below.                      │
└─────────────────────────────────┘


┌─────────────────────────────────┐
│ Have more than six months       │
│ elapsed since (i) purchaser     │
│ complied with notice served by  │          ┌─────────────────────────┐
│ requisite majority of qualifying│          │ Qualifying Tenants have │
│ tenants under s11A, or (ii)     │  If yes  │ lost their rights in    │
│ documents indicating that       │ ────────►│ relation to that        │
│ original disposal has taken     │          │ disposal.               │
│ place, alerting them to         │          └─────────────────────────┘
│ existence of their rights and   │
│ time period for exercise have   │
│ been served on requisite        │
│ majority of qualifying tenants, │
│ or (iii) (where applicable)     │
│ notices under s3A of the        │
│ Landlord & Tenant Act 1985?     │
└─────────────────────────────────┘
         ↙        │         ↘
                If no
```

| If the original disposal consisted of entering into a contract, requisite majority of qualifying tenants may notify L electing for their nominee to step into the shoes of the contracting purchaser—s12A. | If the original disposal consisted of entering into a contract, and no s 12A notice has been served, or if the original disposal did not consist of entering into a contract, requisite majority of qualifying tenants may serve purchase notice on purchaser requiring the purchaser to dispose of its newly acquired interest to nominee—s12B. | If the original disposal consisted of the surrender by L of the tenancy, requisite majority of qualifying tenants may notify purchaser, requiring the purchaser to grant a new tenancy on the same terms to nominee—s12C |

# The Landlord and Tenant Act 1987, Part I (as Amended): Section by Section

**4-01**
1. **Qualifying tenants to have rights of first refusal on disposals by landlord**

1(1) A landlord shall not make a relevant disposal affecting any premises to which at the time of the disposal this Part applies unless—

(a) he has in accordance with section 5 previously served a notice under that section with respect to the disposal on the qualifying tenants of the flats contained in those premises (being a notice by virtue of which rights of first refusal are conferred on those tenants); and

(b) the disposal is made in accordance with the requirements of sections 6 to 10.

**4-02**
**Landlord**
See para 4-29.

**4-03**
**Relevant disposal**
See para 4-47.

## 4-04
## Affecting any premises

It is interesting that section 1(1) of the 1987 Act refers to a disposal "affecting" any premises rather than "of" any premises, which would seem to be more natural. A preliminary point arose in the case of *Dartmouth Court Blackheath Ltd v Berisworth Ltd* [2008] EWHC 350 (Ch) in relation to the identification of "the premises" for the purposes of section 1.

In his judgment, Warren J said, at paras 40–47:

> 40. ... The difference between the parties is most easily illustrated by taking a simple example. Consider a small purpose-built block of flats comprising 9 flats on 3 floors, with 3 flats on each of the 3 floors. The building is in the ownership of a single landlord, L, who has let all of the flats on the ground and first floors on leases under which each tenant is a qualifying tenant. He lets the second floor on short furnished lettings under which the tenants are not qualifying tenants. Suppose that L wishes to dispose of his interest in the second floor by granting a long reversionary lease at a premium but wishes to retain his reversionary interest in the ground and first floors. According to Mr Lidington's approach, section 1 does not apply so that L can grant such a reversionary lease without being required to serve a notice under section 5. The approach of Mr Radevsky, who appears for DCB, leads to precisely the opposite conclusion.
>
> 41. Mr Lidington submits that the focus of sections 1 and 4 is on the property in which the interest of which the landlord is disposing subsists. In the example, this will be the second floor of the building. He identifies that property as the 'premises' with which sections 1(1) and 4(1) are concerned. He says that it is those premises, and those premises alone, which are 'affected' by the disposal which the landlord wishes to make. Having thus identified the premises concerned, it is then necessary to ask whether those premises fall within section 1(2): in the example, they do not because they do not contain any flats occupied by qualifying tenants let alone flats held by qualifying tenants holding more that 50% of the total number of flats in the premises.
>
> 42. In contrast, Mr Radevsky submits that the correct approach is first to identify the relevant premises since only then can one ask whether any particular disposal affects them. In the example, the relevant premises would be the entire block: that is a clear physical unit the entirety of which is in the ownership of L. Having identified the relevant premises, it is then necessary to see whether Part I of the Act

applies. In the example, it clearly would since each of paragraphs (a) to (c) of section 1(2) is satisfied: the block of flats consists of the whole of a building, the building contains 6 flats held by qualifying tenants and those 6 flats exceed 50% of the total number of flats in the block. In contrast with Mr Lidington, Mr Radevsky says that a disposal can affect premises without that disposal needing to relate to each and every part of them. Thus, a disposal of the first floor in the example is a disposal which affects the block as a whole.

43. Mr Lidington submits that Mr Radevsky's approach is not consistent with section 4(1). That provision, it is to be remembered, tells us what the words 'a relevant disposal affecting any premises to which this Part applies' are referring to. That is the phrase which is found (with the addition of 'at the time of the disposal') in section 1(1) itself. The references are to 'the disposal by the landlord of any estate or interest (whether legal or equitable) in any such premises'. Mr Lidington submits that a disposal which 'affects' premises must be of an interest in those same premises. According to this argument, the relevant premises are precisely those, and no more, which the landlord has selected to form the subject matter of his disposal.

44. He also points out that the construction for which Mr Radevsky contends has a rather surprising consequence. Suppose, he says, that a landlord owns a building consisting of retail shops on the ground floor and flats on a number of floors above. On the construction for which Mr Radevsky contends, and assuming the flats are all subject to leases under which the lessees are qualifying tenants and that the floor area of the flats exceeds the percentage set out in section 1(3)(b), a disposal of any of the retail premises eg a new letting of a shop which had fallen vacant, would be subject to the provisions of Part I. He says that cannot possibly have been intended by Parliament.

45. For his part, Mr Radevsky submits that, if Mr Lidington is correct, section 4(1)(a) is unnecessary: Part I of the Act could never apply to a single flat in the light of section 1(2)(b) and there would be no need to have an express exclusion of it. Similarly, the inclusion in section 4(1) of the words 'including the disposal of any such estate or interest in any common parts of such premises' would be unnecessary since common parts, taken by themselves, could never be within Part I of the Act. As to common parts, Mr Lidington suggests that the words in the subsection were added out of an abundance of caution, and points to the judicial criticisms of the drafting of the Act in *Denetower Ltd v Toop* [1991] 1 WLR 945 and *Kay Green v Twinsectra Ltd* [1996] 1 WLR 1587. He did not make the same submission in relation to

paragraph (a); but even if he had, it would have been one which I could not accept. It would only be if driven to it by a need to give sensible meaning to Part I that I would reject as otiose what the draftsman appears to have regarded as a matter of importance, devoting a specific paragraph to it.

46. In my judgment, Mr Radevsky's approach is to be preferred. The relevant premises are to be ascertained in an objective way disregarding the disposal concerned; many factors may come into play in determining the extent of the relevant premises. The provisions of section 1(2) are then applied to those relevant premises to see whether they are premises to which Part I applies. On this construction of the Act, section 4(1)(a) plays a real and important part, permitting the landlord to lease a single flat without qualifying tenants having any right of pre-emption. Further, an intended disposal of common parts would also fall within Part I. That makes sense: it would be odd if a landlord could dispose of common parts separately from any flat without the qualifying tenants being able to acquire them. After all, one purpose of Part I is surely to bring within the control of the qualifying tenants the premises of which their flats form part, a purpose which could easily be defeated if a landlord could dispose of common parts to whomsoever he wished.

47. On this approach, I regard a disposal of part of relevant premises as being a disposal affecting those premises within the meaning of section 4(1). In the context of Part I, it makes perfectly good sense, in my judgment to speak of the disposal of an interest in part of a property as being a disposal of an interest in such property. In the example, the grant of the reversionary lease in the first floor would be 'the disposal of an interest (whether legal or equitable)' in the block: the reversionary lease would not need to relate to the whole block in order to fall within the sub-section.

Note that the use of the word "affecting" rather than "of" has led to the conclusion that a disposal of part of a property can be a disposal that *affects* the whole of the property, and thus all the qualifying tenants of the property must be offered the opportunity to take the disposal, despite the fact that the disposal relates to premises in which only a limited number of such tenants (or, indeed, none of them) have flats.

## 4-05
## Premises

Section 1(2) of the 1987 Act states that the Act applies to premises if all of the following apply:

- they consist of the whole or part of a building
- they contain two or more flats held by qualifying tenants
- the number of such flats exceeds 50% of the total number of flats in the premises.

See para 4-17 in relation to "building", para 4-19 in relation to "flat", para 4-18 in relation to "two or more flats", para 4-21 in relation to "exceeds 50%", para 4-23 in relation to premises used (wholly or partly) otherwise than for residential purposes and para 4-37 in relation to "qualifying tenants".

Sections 62(4) and 55 provide that the 1987 Act applies to England, Wales and the Isles of Scilly only.

## 4-06
## To which at the time of the disposal

Premises may fall within the 1987 Act at one time but fall outside of the Act at another time. For example, the proportion of qualifying tenants to non-qualifying tenants may change, or the amount of commercial space may alter. Some changes may be within the landlord's control, such as letting a vacant flat to an assured shorthold tenant rather than a long leaseholder, but others may be outside of the landlord's control, such as the death or departure of a Rent Act tenant. In the case of a new development, there will logically be a period of time *before* the premises constitute "a building" containing "two or more flats"; and at that stage, presumably the 1987 Act will not apply to those premises (although there is no case law on this point).

The timing issue is particularly significant because of the various steps that the landlord can take at a time when the premises do not fall within the 1987 Act in order to allow the landlord to deal freely with the premises once they do fall within the Act: see Chapter 6, paras 6-05–6.07 and para 6-09.

It is interesting that section 1(1) refers to the premises falling within the Act at the time of the disposal, since the landlord will have to serve section 5 notices and comply with the other requirements of the Act before the disposal is made. The landlord may not necessarily know whether or not the premises will actually fall within the Act at the time of the disposal.

## 4-07
## In accordance with section 5
See para 4-81.

## 4-08
## Previously served
Section 1(1) of the 1987 Act, which requires a notice to be "previously served" before the landlord may make a disposal, contrasts with section 5(1), which requires the landlord to serve a notice where the landlord "proposes to make a relevant disposal". Section 1(1) appears to provide that, so long as notice is served *before* the relevant disposal is made, the landlord will have complied with the landlord's duties under the 1987 Act; whereas section 5(1) implies that the landlord has a duty to serve notice *as soon as* the landlord proposes to dispose.

Note, however, that section 10A(1) provides that the landlord will commit a criminal offence if a disposal is made without "having first complied with the requirements of section 5"—does that mean that no criminal offence is committed so long as notice is served before the disposal, or do the strict requirements of section 5(1) mean that a criminal offence may potentially be committed simply because the notice is served before the disposal is made but long after the landlord first "proposes to dispose"?

See para 4-79 for the meaning of "proposes to dispose".

## 4-09
## Served
See para 4-82.

## 4-10
## Notice
See para 4-83.

## 4-11
## Qualifying tenants
See para 4-37.

## 4-12
### Flat
See para 4-19.

## 4-13
### Rights of first refusal
The rights conferred on the qualifying tenants are rights of first refusal, ie pre-emption rights. As such they offer the tenants only two choices: to accept the offer made (subject to contract) by the landlord, or not to accept that offer. There is no power for the tenants to negotiate the terms of the offer with the landlord (unless, of course, the landlord chooses to negotiate with the tenants), or indeed to argue that the price at which the landlord proposes to dispose of the property is excessive.

## 4-14
### Conferred
There would appear to be a conflict between the preamble to the 1987 Act, which refers to the Act as being "an Act to confer on tenants of flats rights with respect to the acquisition by them of their landlord's reversion", and the reference in section 1(1)(a) to the rights of first refusal in fact only being conferred by virtue of the landlord serving section 5 notices. (Indeed, there are other conflicts between the Act itself and its preamble, including the fact that the rights of first refusal do not necessarily relate to the acquisition of the landlord's reversion but to the acquisition of the interest of which the landlord is proposing to dispose, which may or may not be the landlord's reversionary interest.)

This point is relevant because section 5(5) of the 1987 Act, in stating that section 5 has been complied with where the landlord omits to serve notice on one qualifying tenant (where there are fewer than 10 qualifying tenants), or omits to serve notices on 10% of the qualifying tenants (where there are 10 or more qualifying tenants), effectively permits the landlord to leave certain tenants out of the service process. This means, according to section 1(1)(a), that those tenants who have not received notices, do not have rights under the 1987 Act. Potentially, therefore, a landlord could deliberately omit to serve a notice on a tenant whom the landlord believes is going to respond in a particular way and that tenant would have no right to insist on being served with a notice because the tenant has no rights under the 1987 Act *until* the tenant receives such a notice.

## 4-15
### In accordance with the requirements of sections 6–10
Section 6 of the 1987 Act prohibits the landlord from making a proposed disposal, except to the person nominated by the tenants to take the disposal, during the period in which the section 5 offer may be accepted or, if the section 5 offer is accepted by a sufficient number of tenants, during the period in which a nomination may be made. Similar restrictions relate to the period after which a nomination is made.

As a result of the requirement to comply with sections 6–10, the landlord is not free to proceed with the disposal just because the landlord has received a sufficient number of negative responses from the tenants at any stage during the process laid down by the 1987 Act, for two reasons: first, because the tenants may potentially change their responses at any time up to the end of the relevant period, and second, because section 6 specifically prohibits the making of the disposal, except to the nominated person, during the specified period.

## 4-16
(2) Subject to subsections (3) and (4), this Part applies to premises if —

   (a) they consist of the whole or part of a building; and
   (b) they contain two or more flats held by qualifying tenants; and
   (c) the number of flats held by such tenants exceeds 50 per cent of the total number of flats contained in the premises.

## 4-17
### Building
"Building" is not defined under the 1987 Act. In the case of *Moir* v *Williams* [1892] 1 QB 264, Lord Esher MR defined a "building" as "an inclosure of brick or stone covered in by a roof unless the statute imports a different meaning of the word", which suggests that a structure without a roof cannot be a building—see Chapter 6, para 6-04.

Two particularly difficult areas to deal with, in the context of the Act, are terraces (where the issue is whether it is the terrace as a whole that qualifies as the "building", or whether each individual property within the terrace is itself "a building") and building estates (where the issue is whether each individual block of flats is a separate "building" or whether the estate as a whole can be regarded as a single "building scheme").

- Relevant factors that may assist in deciding which is "the building" in the case of a terrace include:

  — whether the individual flats cross the original party walls of the individual properties within the terrace — if they do, the terrace itself will certainly be the "building" for 1987 Act purposes, even if the party wall(s) is/are only breached on one floor of the terrace, such as in the case of a penthouse flat conversion that extends across the top floors of a number of the original houses within the terrace

  — whether services are shared by the flats within each individual property. If, for example, there is one central heating system for each property, with one shared boiler within that property, then the property is likely to be the "building" for the purposes of the 1987 Act.

Where there is no overlap beyond the original party walls, and there is a separate arrangement of services and service charge provision for each individual property, the likelihood is that the property will be the "building" for the purposes of the Act.

Where there is doubt on the point, a landlord may be well advised to treat the smaller unit as "the building" for the purposes of the Act. The reasoning behind this principle is that a court is likely to have greater sympathy for the tenants seeking to acquire their homes than for the landlord. The smaller the unit that constitutes "the building", the easier it will be for the qualifying tenants to act together to exercise their rights. Where, however, the landlord insists on regarding a whole terrace, comprising perhaps more than 100 flats, as a single "building", rather than splitting that terrace into individual properties of maybe six or eight flats each, the court may regard the landlord as having deliberately made it more difficult for the tenants to exercise their rights under the Act by increasing the number of tenants who need to act together.

There is only one case that deals with the issue of the terrace, and this is the case of *Saga Properties Ltd* v *Palmeira Square Nos 2-6 Ltd* [1995] 15 EG 109. In this case, the Lands Tribunal considered what constituted "the building" for the purposes of the 1987 Act in relation to a terrace of houses, where all of the houses had been altered to provide flats behind the original façade, and where the flats overlapped the line of the original party walls (in the case of four out of five of the original houses). Although the Tribunal decided that it did not have jurisdiction to deal with this point, it went on to say that, if it had had the necessary jurisdiction, it would have held that the four houses containing overlapping flats together constituted a "building" for the purposes of the Act, with the fifth house being a separate building.

- The difficulty with an estate containing a number of separate blocks of flats, all set within a landscape, and all sharing amenities such as private roads, amenity land, etc is to decide whether each separate block, or the estate itself, is the "building" for the purposes of the Act. If each separate block is in fact the "building" for Act purposes, the shared amenities may be so interlinked that it may be virtually impossible to dispose of the reversion to the flat leases.

    There have been a number of cases on this issue, including two Court of Appeal decisions, the first being *Denetower Ltd* v *Toop* [1991] EGCS39 (where there were two blocks of flats facing on to a road, divided by a roadway which led to the shared garages, and the Court held that "building" should be interpreted to include the appurtenances of the building, so that the whole estate was the "building" for Act purposes) and the second being *Kay-Green* v *Twinsectra Ltd* [1996] 2 EGLR 43 (where there were four separate blocks in a shared landscape (albeit built over a period of some four centuries), and the Court held that each of the blocks was a separate "building" for Act purposes).

    Subsequently, however, the High Court held in *Long Acre Securities Ltd* v *Karet* [2004] EWHC 442 (Ch) that "building" could mean "building scheme" for the purposes of the Act. Building scheme meant any building or buildings built as part of a single development at the same time, provided that the occupants of the qualifying flats in each of the buildings share the use of the same appurtenant premises. This is certainly the practical solution to the problem of estates, and the decision in *Kay-Green* v *Twinsectra* can be distinguished on the grounds that the separate blocks had not been built as part of a single development scheme.

- The case of *Dartmouth Court Blackheath Ltd* v *Berisworth Ltd* is also helpful in determining the extent of a "building" for the purposes of the 1987 Act. In this case, the landlord of a block of flats made two disposals without complying with the requirements of the 1987 Act.

The first disposal was a transfer of three garages, the equipment room, caretaker's office and electricity sub-station, together with certain easements and appurtenant rights. The second disposal related to the grant of a lease of the airspace above the roof, the light well, basement rooms and a small area adjoining the wall at the rear of the block of flats over the disused coal chute. The tenants sought to acquire the subject matter of each disposal from the disponee.

First, it was necessary for the Court to determine what was meant by "the premises" in sections 1(1) and 4(1) of the 1987 Act, since the Act prohibits "a relevant disposal affecting any premises" without compliance with the requirements of the Act. Warren J held that the relevant premises must be ascertained in an objective way, disregarding the disposal concerned. The following provisions of the Act must then be applied to those relevant premises to see whether they are caught by the Act.

Each of the disposals was then considered. In the case of the first disposal, the garages, equipment room, caretaker's office and electricity sub-station were all situated within a separate building to the rear of the block of flats itself. The tenants had no rights over or interest in any of this accommodation, the accommodation did not form part of the block of flats (being a separate building) and was not appurtenant to the block of flats, and the Court therefore held that the first disposal was not caught by the 1987 Act.

In the case of the second disposal, however, the premises demised by the lease were held either to be part of the block of flats, or to be appurtenant thereto. The most interesting question related to the airspace above the roof: the tenants had no rights to go on the roof, and that suggested that the airspace was not "appurtenant". However, Warren J stated that the airspace, at least the height of the chimneys, "is an essential part of the space over which any owner of the [block of flats] with repairing obligations would need to have adequate rights of access" and accordingly should be regarded as appurtenant to the building if not actually part of it. "If that is wrong", he continued, "I would conclude that the airspace above the roof to that height is a 'common part' being part of the exterior of the building". The second disposal was therefore caught by the 1987 Act.

## 4-18
## Two or more flats
A building that contains a single flat will never fall within the 1987 Act, but may fall within the Leasehold Reform Act 1967 (as amended): see para 2.12.

## 4-19
## Flat
"Flat" is defined by section 60(1) of the 1987 Act as a separate set of premises (whether or not on the same floor) which forms part of a building, is divided horizontally from some other part of that building and is constructed or adapted for the purposes of a dwelling. There is no case law on the definition of "flat" as meaning a separate set of premises, but arguably a dwelling unit that shares with another unit essential amenities, such as a kitchen or bathroom, will not fall within the definition of "flat" for the purposes of the Act.

"Dwelling" is defined in section 60(1) as:

> a building or part of a building occupied or intended to be occupied as a separate dwelling, together with any yard, garden, outhouses and appurtenances belonging to it or usually enjoyed with it.

It should be noted that, where a building is converted into maisonettes or duplexes in such a way that no unit lies above or below any other unit (as in the case of a house divided into two duplexes, one comprising the whole of the west side of the house and the other the whole of the east side of the house), neither of these units will be a "flat" for the purposes of the Act.

## 4-20
## Qualifying tenants
See para 4-37.

## 4-21
## Exceeds 50%
More than half of the flats in the building must be let to qualifying tenants: section 1(2)(c) of the 1987 Act. This means that in the case of a building containing only two flats, both must be let to qualifying tenants. In the case of a building containing an even number of flats,

*Section by Section*

one more than half of the flats must be let to qualifying tenants, as the following table shows:

| Number of flats in building | Number of flats required to be let to qualifying tenants if the building is to fall within the Act |
|---|---|
| 2 | 2 |
| 3 | 2 |
| 4 | 3 |
| 5 | 3 |
| 6 | 4 |
| 7 | 4 |
| 8 | 5 |
| 9 | 5 |
| 10 | 6 |

### 4-22

(3) This Part does not apply to premises falling within subsection (2) if—

(a) any part or parts of the premises is or are occupied or intended to be occupied otherwise than for residential purposes; and

(b) the internal floor area of that part or those parts (taken together) exceeds 50 per cent of the internal floor area of the premises (taken as a whole);

and for the purposes of this subsection the internal floor area of any common parts shall be disregarded.

### 4-23
### Otherwise than for residential purposes

Section 1(3) of the 1987 Act states that the Act does not apply to premises if any part or parts of the premises is or are occupied (there is no definition of "occupied") or intended to be occupied other than for residential purposes and the internal floor area of that part or parts, taken together, exceeds 50% of the internal floor area of the premises as a whole ("internal floor area" is not defined). Common parts, defined by section 60(1) as "the structure and exterior of [the] building or part and any common facilities within it", are ignored for the purposes of calculating the internal floor area.

In the case of a mixed use building, therefore, the Act will only apply if the internal floor area of the residential part represents at least 50% of the internal floor area of the premises as a whole. A building that contains 50.01% commercial will therefore fall outside of the Act, whereas a building that contains 50% commercial will fall within the Act. The fact that the commercial part of the building is not in fact occupied appears to make no difference, so long as this space is "intended to be occupied" for commercial purposes.

Where the commercial part of a building appears to represent approximately half of the space within the building, it may be worth the landlord's while having a detailed professional survey of the building carried out, as the extent of the common parts within the residential part of the building may be enough to tip the balance and take the building outside of the Act.

There is no definition of "residential purposes" under the 1987 Act and no cases on this. However, obvious areas for possible dispute would include flats used partly for business purposes (such as a doctor's surgery or consulting rooms), as well as commercial premises used partly for living in (such as a restaurant with accommodation for the staff to live in when off-duty: see *Gaingold Ltd v WHRA RTM Company* [2006] 03 EG 122, which held that staff accommodation of this kind was used for "residential purposes" and accordingly formed part of the residential part of the building for the purposes of the 1993 Act).

### 4-24
(4) This Part also does not apply to any such premises at a time when the interest of the landlord in the premises is held by an exempt landlord or a resident landlord.

### 4-25
**Exempt landlord/resident landlord**
See para 4-29.

### 4-26
(5) The Secretary of State may by order substitute for the percentage for the time being specified in subsection (3)(b) such other percentage as is specified in the order.

## 4-27

No order has yet been made by the Secretary of State under this subsection. A statutory instrument that amends this percentage will be subject to annulment pursuant to a resolution of either House of Parliament: section 53(2) of the 1987 Act.

## 4-28

2. Landlords for the purposes of Part I

2(1) Subject to subsection (2) and section 4(1A), a person is for the purposes of this Part the landlord in relation to any premises consisting of the whole or part of a building if he is—

(a) the immediate landlord of the qualifying tenants of the flats contained in those premises, or

(b) where any of those tenants is a statutory tenant, the person who, apart from the statutory tenancy, would be entitled to possession of the flat in question.

## 4-29
### Landlord

- Section 20(1) of the 1987 Act states that "landlord" should be "construed in accordance with Section 2". Section 2(1) defines "landlord" (subject to subsection (2) and section 4(1A)) as the immediate landlord of the qualifying tenants or, in the case of a statutory tenancy (see Chapter 4, para 4-32), the person who would be entitled to possession of the flat if it were not for the tenancy. It is difficult to see why the Act draws a distinction between these two types of landlord, as a tenant's immediate landlord will always, by definition, be entitled to possession of the tenant's flat if it were not for the tenant's tenancy of it.

- Section 2(2) states that, where the immediate landlord of the qualifying tenants is also a tenant, and the landlord's lease is either for less than seven years, or contains a right for the superior landlord to end the lease (presumably this means a specific landlord's break right rather than a mere right of forfeiture for breach of the lease by the tenant) within the first seven years of the term, then the superior landlord is the "landlord" of the qualifying tenants for the purposes of the Act. As a result, any disposal ma by the superior landlord is capable of being a "relevant disposal" but any disposal made by the immediate landlord does not fall within the ambit of the Act and the qualifying tenants have no rights in relation to it.

It does not appear to be relevant that the intermediate tenancy, although capable of being terminated within the first seven years, was not in fact so terminated; indeed, whether or not the superior landlord's right to terminate remains effective as at the date of the proposed disposal also appears to be irrelevant for this purpose. It seems unlikely that the draftsman intended so wide an interpretation of this section; if the first seven years of the term have elapsed at the relevant date, without the superior landlord exercising a break right, then the intermediate tenancy should surely not be ignored, although this is required by the strict wording of the statute. It seems harsh that the qualifying tenants of a building could be prevented from acquiring, say, a 999-year lease of that building simply because that lease contained a landlord's break right that has long since expired without being exercised.

- Section 4(1A) of the Act states that a mortgagee who exercises a power of sale or leasing under a mortgage will treated as a landlord for the purposes of the Act, whether the sale or leasing is done in the name of the landlord mortgagor or not.

- Note that where an agreement for lease exists, the parties to that agreement will be a "landlord" and a "tenant" respectively for the purposes of the Act. Section 59(1) states that "lease" includes an agreement for lease, and section 59(2) states that "landlord" and "tenant" shall be construed accordingly.

- Note that a landlord who is an "exempt landlord" or a "resident landlord" is still a landlord for the purposes of the Act, but the Act does not apply to premises held by an exempt or resident landlord: section 1(4) of the Act.

    Section 58(1) defines "exempt landlord" by way of a list of public sector bodies comprising:

    (a) a district, county, county borough or London borough council, the Common Council of the City of London, the London Fire and Emergency Planning Authority, the Council of the Isles of Scilly, a police authority established under section 3 of The Police Act 1996 or a joint authority established by Part IV of the Local Government Act 1985;

(b) the Commission for the New Towns or a development corporation established by an order made (or having effect as if made) under The New Towns Act 1981;

(c) an urban development corporation within the meaning of Part XVI of the Local Government, Planning and Land Act 1980;

(ca) a housing action trust established under Part III of the Housing Act 1988;

(dd) the Broads Authority;

(de) a National Park authority;

(e) the Housing Corporation;

(f) a housing trust (as defined in section 6 of The Housing Act 1985) which is a charity;

(g) a registered social landlord (as defined in Sections 5(4) and 5(5) of The Housing Act 1985) or a fully mutual housing association (as defined in Sections 1(1) and 1(2) of The Housing Associations Act 1985) which is not a registered social landlord; or

(h) an authority established under section 10 of The Local Government Act 1985 (joint arrangements for waste disposal functions).

Section 58(2) defines "resident landlord" as a landlord who "occupies a flat contained in the premises as his only or principal residence and .... has so occupied such a flat throughout a period of not less than 12 months ending with that time". This applies only where the building is not a "purpose-built block of flats" (defined by section 58(3) as "a building which contained as constructed, and contains, two or more flats").

The Crown can never be a landlord for the purposes of the Act. Section 56(1) states that the Act will apply to "a tenancy from the Crown" where there has ceased to be a Crown interest in the land. Section 56(4) defines "a tenancy from the Crown" as "a tenancy of land in which there is, or has during the subsistence of the tenancy been, a Crown interest superior to the tenancy", and defines "Crown interest" as an interest comprised in the Crown Estate or belonging to Her Majesty in right of the Duchy of Lancaster or belonging to the Duchy of Cornwall or belonging to a government department or held on behalf of Her Majesty for the purposes of a government department. The wording of section 56(4) implies that, even if the Crown ceases to be the immediate landlord of the qualifying tenants, the Act will still not apply to a disposal by their immediate landlord so long as the Crown holds an interest in the land superior to that of the landlord; however, section 56(3) specifically states that, in such a case, the provisions of the 1987 Act will apply to that landlord.

Where freehold land comes within Crown ownership as a result of a disclaimer by the previous owner's liquidator, the Crown's interest in that land will not be a Crown interest for the purposes of section 56 of the Act: *Scmlla Properties Ltd* v *Gesso Properties (BVI) Ltd* [1995] EGCS52.

## 4-30
## Qualifying tenants
See para 4-37.

## 4-31
## Flats
See para 4-19.

## 4-32
## Statutory tenancy
Section 60(1) of the 1987 Act defines "statutory tenant" and "statutory tenancy" as a statutory tenant or tenancy within the meaning of the Rent Act 1977 or the Rent (Agriculture) Act 1976. A statutory tenancy is the tenancy that arises after a contractual tenancy (called a "protected tenancy") terminates, and that continues so long as the statutory tenant occupies the premises as the tenant's residence.

## 4-33
2(2) Where the person who is, in accordance with subsection (1), the landlord in relation to any such premises for the purposes of this Part ('the immediate landlord') is himself a tenant of those premises under a tenancy which is either —

(a) a tenancy for a term of less than seven years, or
(b) a tenancy for a longer term but terminable within the first seven years at the option of the person who is the landlord under that tenancy ('the superior landlord'),

the superior landlord shall also be regarded as the landlord in relation to those premises for the purposes of this Part and, if the superior landlord is himself a tenant of those premises under a tenancy falling within paragraph (a) or (b) above, the person who is the landlord under that tenancy shall also be so regarded (and so on).

## 4-34

Section 2(2) of the 1987 Act states that, where the immediate landlord of the qualifying tenants is a tenant, and the immediate landlord's lease is either for less than seven years, or contains a right for the superior landlord to end the lease (presumably this means a specific landlord's break right rather than a mere right of forfeiture for breach of the lease by the tenant) within the first seven years of the term, then the superior landlord is the "landlord" of the qualifying tenants for the purposes of the Act. As a result, any disposal made by the superior landlord is capable of being a "relevant disposal" but any disposal made by the immediate landlord does not fall within the ambit of the Act and the qualifying tenants have no rights in relation to it.

It does not appear to be relevant that the intermediate tenancy, although *capable* of being terminated within the first seven years, was not in fact so terminated; indeed, whether or not the superior landlord's right to terminate remains effective as at the date of the proposed disposal, also appears to be irrelevant for this purpose. It seems unlikely that the draftsman intended so wide an interpretation of this section; if the first seven years of the term have elapsed at the relevant date, without the superior landlord exercising the break right, then the intermediate tenancy should surely not be ignored, although this is required by the strict wording of the statute. It seems harsh that the qualifying tenants of a building could be prevented from acquiring, say, a 999-year lease of that building simply because that lease contained a landlord's break right that has long since expired without being exercised.

## 4-35

3. **Qualifying tenants**
3(1) Subject to the following provisions of this section, a person is for the purposes of this Part a qualifying tenant of a flat if he is the tenant

of the flat under a tenancy other than —

(a) a protected shorthold tenancy as defined in section 52 of the Housing Act 1980;
(b) a tenancy to which Part II of the Landlord and Tenant Act 1954 (business tenancies) applies;
(c) a tenancy terminable on the cessation of his employment or
(d) an assured tenancy or assured agricultural occupancy within the meaning of Part I of the Housing Act 1988.

## 4-36
## Tenant

Note that where an agreement for lease exists, the parties to that agreement will be a "landlord" and a "tenant" respectively for the purposes of the 1987 Act. Section 59(1) states that "lease" includes an agreement for lease, and section 59(2) states that "landlord" and "tenant" shall be construed accordingly.

## 4-37
## Qualifying tenant

Section 3 of the 1987 Act defines a "qualifying tenant" as being every tenant of a flat except for the specifically listed exceptions. Accordingly, companies, persons who do not reside in the relevant flat and Rent Act tenants are all "qualifying tenants" for the purposes of the Act. The fact that a Rent Act tenant is a qualifying tenant under the 1987 Act is particularly significant as such a tenant has no rights to participate in collective enfranchisement or the exercise of the right to manage. The 1987 Act is the only legislation that gives such a tenant the possibility of owning a share of the freehold of the tenant's home.

The exceptions are as follows:

- A tenant under a protected shorthold tenancy: section 3(1)(a). A protected shorthold tenancy is defined in section 52 of the Housing Act 1980 and is a very rare type of tenancy that could not be created after 14 January 1989.

- A tenant under a business tenancy to which the Landlord and Tenant Act 1954, Part II applies: section 3(1)(b). A business tenancy in relation to which security of tenure has been excluded pursuant to section 38(4)(a) of the 1954 Act *is* a tenancy to which Part II of the 1954 Act applies; however, Part II of the 1954 Act does not apply to a business tenancy where the tenant is not in occupation of the premises for the purposes of the tenant's business, or a tenancy for a fixed term not exceeding six months, or a tenancy at will. A tenant under such a tenancy *could* therefore be a qualifying tenant for the purposes of the 1987 Act, but for the fact that section 3(1) appears to require that the tenant to be a tenant of a "flat". On the other hand, it seems surprising that a business tenant is specifically excluded from being a qualifying tenant if in fact such tenant has to be a tenant of a flat in the first place. This may be a practical issue in the context of a live/work unit.

- A tenant under a tenancy terminable on the ending of the tenant's employment: section 3(1)(c).

- A tenant under an assured tenancy (including an assured shorthold tenancy) or under an assured agricultural occupancy: section 3(1)(d). Note that a tenancy at a rent of more than £25,000 per annum cannot be an assured tenancy, even if it is described as such, and accordingly a tenant under such a tenancy will be a qualifying tenant.

- A tenant who is the tenant (under one or more leases) of three or more flats in the building (excluding any flat that falls within one of the above four exceptions): section 3(2). Section 3(3) states that, for the purposes of section 3(2), where the tenant is a body corporate, any letting of a flat in the building to an associated company is treated as a letting to that body corporate. Section 20(1) defines "associated company" to mean, in relation to a body corporate, another body corporate which is that body's holding company, or its subsidiary, or a subsidiary of its holding company, in each case within the meaning of section 736 of the Companies Act 1985.

- A tenant whose landlord is also a qualifying tenant of that flat: section 3(4). This means that where both a tenant and a subtenant would otherwise be qualifying tenants, the tenant will in fact be the qualifying tenant, and the subtenant cannot also be a qualifying tenant.

## 4-38
## Flat
See para 4-19.

## 4-39
## Tenancy
Section 59(1) of the 1987 Act provides that "lease" and "tenancy" have the same meaning, and both expressions include a sublease or subtenancy and an agreement for lease (or tenancy or sublease or subtenancy). A tenancy also includes a statutory tenancy within the meaning of the Rent Act 1977 or the Rent (Agriculture) Act 1976: section 60(1).

## 4-40
3(2) A person is not to be regarded as being a qualifying tenant of any flat contained in any particular premises consisting of the whole or part of a building if by virtue of one or more tenancies none of which falls within paragraphs (a) to (d) of subsection (1), he is the tenant not only of the flat in question but also of at least two other flats contained in those premises.

## 4-41
## Qualifying tenant
See para 4-37.

## 4-42
3(3) For the purposes of subsection (2) any tenant of a flat contained in the premises in question who is a body corporate shall be treated as the tenant of any other flat so contained and let to an associated company.

## 4-43
## Qualifying tenant
See para 4-37.

## 4-44
3(4) A tenant of a flat whose landlord is a qualifying tenant of that flat is not to be regarded as being a qualifying tenant of that flat.

## 4-45
## Qualifying tenant
See para 4-37.

## 4-46
4. **Relevant disposals**

4(1) In this Part references to a relevant disposal affecting any premises to which this Part applies are references to the disposal by the landlord of any estate or interest (whether legal or equitable) in any such premises, including the disposal of any such estate or interest in any common parts of any such premises but excluding —

(a) the grant of any tenancy under which the demised premises consist of a single flat (whether with or without any appurtenant premises); and
(b) any of the disposals falling within subsection (2).

## 4-47
## Relevant disposal

- "Relevant disposal" means any disposal by the landlord of any legal or equitable estate or interest in any premises to which the Act applies (including the surrender of a tenancy (but not where this is made in pursuance of a covenant, condition or agreement contained in the tenancy), or grant of an option or right of pre-emption (section 4(3)(a)), the disposal of an interest in the common parts (section 4(1)) and a sale by a mortgagee exercising power of sale (section 4(1A))) will be caught by the Act, apart from the exceptions listed in sections 4(1) and 4(2) of the Act. "Disposal" is itself defined by section 20(1) of the Act as "construed in accordance with Section 4(3) and Section 4A" of the Act.

- Section 4A(1) of the Act expressly states that the Act applies to "a contract to create or transfer an estate or interest in land, whether conditional or unconditional, and whether or not enforceable by specific performance": this provision expressly overrules the decision of the Court of Appeal in the case of *Mainwaring* v *Henry Smith's Charity Trustees* [1996] 2 EGLR 25, when it was held that an exchange of contracts did not constitute a disposal for the purposes of the 1987 Act (prior to its amendment by the Housing Act 1996). It would be possible for a landlord to enter into a contract with a third party relating to the sale of the landlord's interest in the premises, but for the sale and purchase pursuant to that contract to be subject to a condition precedent, namely that the qualifying tenants fail to exercise their rights under the 1987 Act. Such a contract, if properly drafted, would not constitute "a disposal of any legal or equitable estate or interest" in the relevant premises, but exchange of such a contract would still be caught by the Act by virtue of the provisions of section 4A(1).

- It would be possible for a landlord and a third party each to sign their part of a contract for the sale and purchase of the landlord's interest in the premises and then to hand over each part to the other's legal representative (without effecting an exchange of contracts). Each part of the contract could then be held by the legal representative subject to an escrow (using solicitors' undertakings) whereby the contract could only be dated and must in fact then be dated (in which case exchange of contracts would be deemed to take place) after the qualifying tenants have failed to exercise their rights under the 1987 Act, so that the landlord is then free to dispose to the third party purchaser. A contract held in escrow in this way would not be a relevant disposal, on the strict wording of the Act, and would appear to afford to the landlord and the third party the only lawful means of binding each other to the transaction whilst the result of section 5 or section 18 notices is awaited. This would be particularly important in the case of a volatile market, where each party needs to know that the other party is bound to the transaction at the previously agreed price.

- Section 4A(2) states that the provisions of the 1987 Act apply to an assignment of rights under a contract and section 4A(3) states that the Act also applies to a contract to make an assignment of rights under a contract. Further, section 4A(2)(c) states that references to the landlord shall be construed as references to the assignor. Section 4A(4), however, states that nothing in this section affects the operation of the 1987 Act relating to options or rights of pre-emption.

    See comments in para 4-74 in relation to the effect of these provisions.

- Note that section 11(3) makes it clear that a disposal for no consideration is still a relevant disposal: see also the case of

*Okonedo* v *Kirby* [2006] EW Lands LRX_15_2006, in which the landlords had transferred the freehold of a building to their friend and cousin as a gift for no consideration and without service of offer notices, and the Court of Appeal (Lloyd LJ), upholding the decision of the Lands Tribunal, held that the transferee must pass the building to the nominated purchaser for no consideration. Article 1 of the First Protocol of the European Convention on Human Rights did not assist the transferee.

- The exceptions listed in sections 4(1) and 4(2) of the Act are:

  — the grant of a tenancy of a single flat (including any appurtenant premises): section 4(1). Section 4(4) defines "appurtenant premises", in relation to any flat, as "any yard, garden, outhouse or appurtenance (not being a common part of the building containing the flat) which belongs to, or is usually enjoyed with, the flat": see para 4-65

  Note that the grant of a tenancy of a single commercial unit is not specifically exempted from the 1987 Act, although it is not clear how the requirements of the 1987 Act would interact with the requirements of Part II of the Landlord and Tenant Act 1954. The issue of a letting of a single shop was addressed in the case of *Dartmouth Court Blackheath Ltd* v *Berisworth Ltd*, when Warren J rejected an argument by Counsel that would have avoided such a letting falling foul of the Act: see paras 44–46 of that judgment, set out in para 4-04 above.

  — the disposal of any interest of a beneficiary in settled land: section 4(2)(a)(i)

  — the disposal of any incorporeal hereditament (this includes a rentcharge, annuity or corrody, advowson, tithe, easement, profit a prendre, title of honour, office or franchise: section 4(2)(a)(iii)

  — the disposal by way of security for a loan (ie the creation of a mortgage or charge): section 4(2)(aa)

  — a disposal to a trustee in bankruptcy or the liquidator of a company: section 4(2)(b)

- various disposals pursuant to a court order made in specified matrimonial, family or inheritance proceedings: section 4(2)(c)

- a disposal pursuant to a compulsory purchase order or an agreement entered into by way of compromise to avoid a compulsory purchase order being made or carried into effect: section 4(2)(d)

- a disposal by way of collective enfranchisement or lease renewal under the Leasehold Reform, Housing and Urban Development Act 1993: section 4(2)(da)

- a disposal by way of gift to a member of the landlord's family or to a charity: section 4(2)(e)

- a disposal of functional land by one charity to a second charity which intends to use that land as functional land: section 4(2)(f). Note that land held by a charity for investment or commercial purposes may not constitute "functional land" under section 60(1) of the Act, which defines "functional land" as "land occupied by the charity, or by trustees for it, and wholly or mainly used for charitable purposes"

- a disposal of land held on trust in connection with the appointment or discharge of a trustee: section 4(2)(g)

- a disposal by two or more members of a family to fewer of their number, or to a different combination of family members (to include at least one of the transferors): section 4(2)(h). Sections 4(5) and 4(6) define in greater detail what is meant by being a member of another's family, and in particular what is meant by "spouse", "parent", "grandparent", "child", "grandchild", "brother", "sister", "uncle", "aunt", "nephew" and "niece"

- a disposal in pursuance of a contract, option or right of pre-emption binding on the landlord (except as provided by section 8D of the Act): section 4(2)(i). Note that the exchange of contracts or grant of the option or right of pre-emption will, however, be a relevant disposal

- the surrender of a tenancy pursuant to a covenant, condition or agreement contained in that tenancy: section 4(2)(j)

— a disposal to the Crown: section 4(2)(k)

— a disposal by a body corporate to a company which has been an associated company of that body for at least two years: section 4(2)(l). Section 20(1) defines "associated company" to mean, in relation to a body corporate, another body corporate which is that body's holding company, or its subsidiary, or a subsidiary of its holding company, in each case within the meaning of section 736 of the Companies Act 1985. There is thus a discrepancy between section 4(2)(l), which refers to a disposal to a company, and section 20(1), which refers throughout to bodies corporate. It is not clear whether or not a disposal between two associated bodies corporate, not being companies (such as two limited liability partnerships) would satisfy the requirements of the Act

— a disposal under the terms of a will or the law relating to intestacy: section 4(3)(b).

## 4-48
### Affecting any premises
See para 4-04.

## 4-49
### Landlord
See para 4-29.

## 4-50
### Common parts
"Common parts" is defined by section 60(1) as "the structure and exterior of [the] building or part and any common facilities within it". Note that the grant of a lease of the common parts of a building to a management company, even where the shares in that company are to be held by the tenants of the flats, is a relevant disposal under the 1987 Act, and there is no exception covering this.

## 4-51
### Flat
See para 4-19.

## 4-52
### Appurtenant premises
See para 4-65.

## 4-53
> 4(1A) Where an estate or interest of the landlord has been mortgaged, the reference in subsection (1) above to the disposal of an estate or interest by the landlord includes a reference to its disposal by the mortgagee in exercise of a power of sale or leasing, whether or not the disposal is made in the name of the landlord; and, in relation to such a proposed disposal by the mortgagee, any reference in the following provisions of this Part to the landlord shall be construed as a reference to the mortgagee.

## 4-54
Although the creation of a mortgage or charge as security for a loan is not a relevant disposal and so is not caught by the Act (section 4(2)(aa)), a sale by a mortgagee exercising the power of sale will be caught by the Act by virtue of the above section, unless that sale, or other disposal, is not in fact a relevant disposal. If, however, the mortgagee is prepared simply to manage the property following default by the landlord/mortgagor, receiving the rents but not exercising the power of sale, the tenants' rights under the 1987 Act will not be triggered. The mortgagee may even assign the mortgagee's rights under the mortgage to a third party without triggering the tenants' rights — section 4A(2) will not apply, as a mortgage is not a contract. Foreclosure, where the property is vested in the mortgagee by order of the Court, appears not to be a relevant disposal, as there is no disposal as such, merely a vesting. There may be possible ways of avoiding the 1987 Act by using a mortgage: see Chapter 6 para 6-11 below.

## 4-55
### Following
Section 4(1A) provides that "landlord" includes "mortgagee" in relation to "the following provisions" of Part I of the 1987 Act. Section 1(4), which provides that where the landlord is an exempt landlord or resident landlord, the 1987 Act does not apply to the premises, comes *before* section 4(1A) in the 1987 Act. This suggests that a mortgagee cannot take advantage of the exempt landlord and resident landlord provisions when disposing of the premises.

For example, a public body falling within the definition of "exempt landlord" takes a charge over privately-owned residential premises in order to secure a loan: the loan provisions are not complied with and the body decides to sell the premises in order to recover the loan. The original taking of the charge was not a relevant disposal and so did not confer rights of first refusal on the tenants — see section 4(2)(aa) above. Normally, a disposal of premises by the public body would be outside the 1987 Act, because it is an exempt landlord. However, in this case, the body is disposing in its capacity as mortgagee and hence the residential tenants will enjoy rights of first refusal.

In some ways, the above example is not particularly surprising; the landlord is not an exempt landlord so why should the residential tenants not enjoy their rights of first refusal despite the fact that the disposal is made by a mortgagee? Where, however, the landlord is an exempt or resident landlord, and the disposal is made by a mortgagee that would otherwise fall within the exempt landlord or resident landlord definition, the tenants appear to enjoy greater rights under the 1987 Act where the disposal is made by the mortgagee than where the disposal is made by the landlord.

## 4-56

4(2) The disposals referred to in subsection (1)(b) are—

(a) a disposal of —

(i) any interest of a beneficiary in settled land within the meaning of the Settled Land Act 1925, or
(ii) ...
(iii) any incorporeal hereditament;

(aa) a disposal by way of security for a loan;

(b) a disposal to a trustee in bankruptcy or to the liquidator of a company;

(c) a disposal in pursuance of an order made under -

(i) section 24 of the Matrimonial Causes Act 1973 (property adjustment orders in connection with matrimonial proceedings),
(ii) section 24A of the Matrimonial Causes Act 1973 (orders for the sale of property in connection with matrimonial

proceedings) where the order includes provision requiring the property concerned to be offered for sale to a person or class of persons specified in the order,

(iii) section 2 of the Inheritance (Provision for Family and Dependants) Act 1975 (orders as to financial provision to be made from estate),

(iv) section 17(1) of the Matrimonial and Family Proceedings Act 1984 (property adjustment orders after overseas divorce, &c.),

(v) section 17(2) of the Matrimonial and Family Proceedings Act 1984 (orders for the sale of property after overseas divorce, &c.) where the order includes provision requiring the property concerned to be offered for sale to a person or class of persons specified in the order,

(vi) paragraph 1 of Schedule 1 to the Children Act 1989 (orders for financial relief against parents),

(vii) Part 2 of Schedule 5, or paragraph 9(2) or (3) of Schedule 7, to the Civil Partnership Act 2004 (property adjustment orders in connection with civil partnership proceedings or after overseas dissolution of a civil partnership, etc), or

(viii) Part 3 of Schedule 5, or paragraph 9(4) of Schedule 7, to the Civil Partnership Act 2004 (orders for the sale of property in connection with civil partnership proceedings or after overseas dissolution of a civil partnership, etc) where the order includes provision requiring the property concerned to be offered for sale to a person or class of persons specified in the order.

(d) a disposal in pursuance of a compulsory purchase order or in pursuance of an agreement entered into in circumstances where, but for the agreement, such an order would have been made or (as the case may be) carried into effect;

(da) a disposal of any freehold or leasehold interest in pursuance of Chapter I of Part I of the Leasehold Reform, Housing and Urban Development Act 1993;

(e) a disposal by way of gift to a member of the landlord's family or to a charity;

(f) a disposal by one charity to another of an estate or interest in land which prior to the disposal is functional land of the first-mentioned charity and which is intended to be functional land of the other charity once the disposal is made;

(g) a disposal consisting of the transfer of an estate or interest held on trust for any person where the disposal is made in connection with the appointment of a new trustee or in connection with the discharge of any trustee;

(h) a disposal consisting of a transfer by two or more persons who are members of the same family either—

   (i) to fewer of their number, or
   (ii) to a different combination of members of the family (but one that includes at least one of the transferors);

(i) a disposal in pursuance of a contract, option or right of pre-emption binding on the landlord (except as provided by section 8D (application of sections 11 to 17 to disposal in pursuance of option or right of pre-emption));

(j) a disposal consisting of the surrender of a tenancy in pursuance of any covenant, condition or agreement contained in it;

(k) a disposal to the Crown; and

(l) a disposal by a body corporate to a company which has been an associated company of that body for at least two years.

## 4-57
## Incorporeal hereditament

An incorporeal hereditament includes a rentcharge, annuity or corrody, advowson, tithe, easement, profit a prendre, title of honour, office or franchise.

## 4-58
## Disposal by way of security for a loan

Note that the creation of a legal charge or equitable charge over premises is not necessarily outwith the definition of "relevant disposal", as a charge that is not made by way of security for a loan (eg a charge made to secure overage) will be a relevant disposal.

## 4-59
## Member of family

See para 4-69 and para 4-70.

## 4-60
## Charity
Section 60(1) defines "charity" to mean a charity within the meaning of the Charities Act 1993 and "functional land" to mean land occupied by the charity or by trustees for it, and wholly or mainly used for charitable purposes.

## 4-61
## Disposal in pursuance of a contract, option or right of pre-emption binding on the landlord
Note that the exchange of contracts or grant of the option or right of pre-emption will, however, be a relevant disposal.

## 4-62
## Body corporate/associated company
Section 20(1) defines "associated company" to mean, in relation to a body corporate, another body corporate which is that body's holding company, or its subsidiary, or a subsidiary of its holding company, in each case within the meaning of section 736 of the Companies Act 1985. There is thus a discrepancy between section 4(2)(l), which refers to a disposal to a company, and section 20(1), which refers throughout to bodies corporate. It is not clear whether or not a disposal between two associated bodies corporate, not being companies (such as two limited liability partnerships) would satisfy the requirements of the Act.

## 4-63
> 4(3) In this Part 'disposal' means a disposal whether by the creation or the transfer of an estate or interest and—
>
> (a) includes the surrender of a tenancy and the grant of an option or right of pre-emption, but
> (b) excludes a disposal under the terms of will or under the law relating to intestacy;
>
> and references in this Part to the transferee in connection with a disposal shall be construed accordingly.

## 4-64

4(4) In this section "appurtenant premises", in relation to any flat, means any yard, garden, outhouse or appurtenance (not being a common part of the building containing the flat) which belongs to, or is usually enjoyed with, the flat.

## 4-65
### Appurtenance

The meaning of "appurtenance", in the context of a building rather than a flat, was considered in the case of *Denetower Ltd v Toop*. In that case, two blocks of flats, each consisting of four flats with gardens, were let to tenants. The leases included the relevant part of the garden, and granted rights of way over the roadway running between the blocks and pathways leading to the flats and gardens. At the rear of the flats there were eight garages, accessible by the roadway, but all let on separate leases. Most of the tenants had a long lease of a garage, but not all of them. There was a small piece of unused land over which the tenants had no rights.

Sir Nicolas Browne-Wilkinson, the Vice-Chancellor, held that whilst the gardens, roadways and paths were appurtenances, neither the garages nor the unused land was an appurtenance, because the tenants had no rights over the unused land, and the rights over the garages were granted under separate leases.

In the case of *Dartmouth Court Blackheath Ltd v Berisworth Ltd*, Warren J considered the decision of the Vice-Chancellor and went on to state, at para 53–54:

53. It must be remembered, in applying this decision [i.e. that of *Denetower v Toop*], the Act itself does not expressly include appurtenances as part of the building. What is, and is not, appurtenant in accordance with the judicial gloss which has been placed on the meaning of 'appurtenant' is very much a matter of fact and degree. For example, ordinarily, a garage built next to a single dwelling house and owned and used by the owner and occupier of the house might be thought of as appurtenant to the house and so pass with a transfer of the freehold of the house. It does not follow from that that a row of garages close to a block of flats where the garages are not let to the tenants of the flats as part of their flat tenancies, but rather let separately, is appurtenant (in the sense of the judicial gloss) to the block of flats. Rather, the concept of 'appurtenant' is invoked in this context to avoid what would otherwise be a capricious result in the application of this particular

piece of legislation. That, no doubt, is why the Vice-Chancellor in *Denetower* placed some emphasis on the fact that the gardens and certain other appurtenances were expressly or impliedly included in the demises of the flats to the tenants; they were thus appurtenant to the building as that building was actually used. In contrast, the use of the garages was not by persons *qua* tenants of the flats.

54. This approach reflects the meaning given to appurtenant in section 4(4) expressly referring to 'appurtenant premises' meaning 'any yard, garden, outhouse or appurtenance ... **which belongs to, or is usually enjoyed with, the flat**'.

He then rejected any analogy with the meaning of "appurtenances" for the purposes of the Leasehold Reform Act 1967, and went on to say, in relation to *Dartmouth Court*, at paras 59–70:

59. ... the garages in the Garage Block are not, in my judgment, appurtenant to the Main Building any more than the garages in Denetower were appurtenant to the blocks of flats in that case. The tenants of the garages do not enjoy the use of the garages as part of their enjoyment under their flat tenancies. The garages are held under quite separate leases and not every flat holder has a garage tenancy. Although the garage leases contain restrictions on assignment so that only assignment to a flat holder is permitted, the Landlord is free to lease a garage to whomsoever he wishes. The flat tenants do not, in their capacities as tenants, contribute through the service charge to the maintenance of the Garage Block. The garages are not, I consider, enjoyed in such a way as to make them appurtenant to the Main Building.

60. So far as concerns the mobile phone plant room, there is even less room for arguing that it is appurtenant to the Main Building than is the case in relation to the garages. The plant room houses equipment belonging to a third party in connection with a mast on the roof which serves mobile phone users in the area of coverage of the mast: the mobile phone provision has nothing to do with the use and enjoyment of the Main Building. In my judgment, the plant room is not appurtenant to the Main Building.

61. The caretaker's office and the electricity sub-station give rise to some different considerations. Taking the caretaker's office first, it is said that this is appurtenant to the Main Building because the flat leases require the provision of maintenance staff. It is pursuant to that obligation that the caretaker's services are provided. It does not,

however, follow from that that the office provided for him is appurtenant to the Main Building. The Landlord remains the exclusive owner of the office; the tenants have no interest in it at all. The Landlord is not obliged to the tenants to provide an on-site office for any of the maintenance staff at all. The fact that he has chosen to do so clearly, in my view, does not result in the office which he does provide becoming appurtenant to the Main Building.

62. The position in relation to the electricity sub-station is less clear. It is at least possible, and for the purposes of this case I assume it to be the case, that the sub-station serves only the Property. Even so, I do not consider that the sub-station is appurtenant to the Main Building. The sub-station is let on a long lease to an electricity supplier. It houses the equipment belonging to that supplier. This may be a convenient arrangement, but there is no reason why electricity has to be supplied by means of a sub-station within the Property. The tenants have no right to that particular method of supply and have no interest in the sub-station or the equipment. In my judgment, the sub-station is not appurtenant to the Main Building in the sense required for an appurtenance to be part of building for the purposes of section 1(2).

... ...

67. The question then is whether the airspace above the flat roof and above the mansard roof forms part the building, either as a matter of the ordinary use of language or applying the approach of the Vice-Chancellor in *Denetower* or otherwise. If it is, then the lease of that airspace will be a disposal of part of building and thus fall within section 1(2). But if it is not, the question would then arise whether the airspace, at least to some height above the roof, is part of the 'common parts' of the building.

68. Generally speaking, a transfer of freehold land will carry with it the earth below and the air above that land. That is a general principle not without limits. As the decision of Griffiths J in Bernstein of *Leigh (Baron)* v *Skyviews & General Ltd* [1978] 1 QB 479 shows, an owner's rights in airspace about his land are restricted to such height as is necessary for the ordinary use and enjoyment of the land and structures upon it and above that height he has no greater rights than any other member of the public. Mr Radevsky suggests, on the basis of that decision, that a building includes the airspace above it at least to that extent. In the present case, the building — that is the Main Building — would therefore include the airspace above at least to the height necessary to construct a new floor of flats on the roof space, if not considerably further. Alternatively, such airspace is

appurtenant to the building; and in the further alternative is part of the "common parts" as defined.

69. I do not dissent from the view that a transfer of the Main Building (whether with or without the rest of the Property) would indeed carry with it the airspace above at least to the height described by Griffiths J. But that is because a landowner transferring an area of land together with the building standing on it is presumed to be transferring everything which he owns which will include the airspace which forms part of what he owns. It is an entirely different question whether 'building' in section 1(2) includes the airspace above the bricks, mortar and tiles which comprise the physical building.

70. As to that, the tenants have no rights under their leases to access to the roof. I understand that occasionally in the past, a representative of the tenants has gone onto the roof and into the mansard roof to deal with emergencies such as burst pipes and leaking water tanks. I do not think that that would be sufficient to turn any of the airspace into part of the building if, in the absence of such activity, it would not otherwise be part of the building. But that is not the only consideration. The fact is that the Landlord is under an obligation to keep the structure of the Main Building (including the roofs and chimney stacks) in repair. The Landlord certainly requires access to carry out his obligations in respect of those structures, as well as in relation to the pipes and tanks which are to found on the roof and within the mansard roof. The airspace is not, I accept, appurtenant to the building in the same sense that the gardens were appurtenant to the buildings in *Denetower*, the tenants having no rights over the airspace. But, or so it seems to me, the airspace, at least the height of the chimneys (but see paragraph 73 below), is an essential part of the space over which any owner of the Main Building with repairing obligations would need to have adequate rights of access. At a time when the airspace is actually owned by the owner of the building, I consider that it is correct to regard the airspace up to that height as appurtenant to the building if not actually part of it. To echo the words of the Vice-Chancellor in *Denetower*, it would be to attribute to Parliament an entirely capricious intention if I were to hold that the tenants' rights to purchase did not extend to the airspace above the roof the enjoyment of which is necessary to maintain the structure, including the roof and chimneys, in the state of repair in which the Landlord is obliged to keep it. In my judgment, it is perfectly legitimate meaning of the word 'building' that it includes the airspace necessary to enable maintenance to be carried out.

Finally, Warren J stated that, if he were wrong about the airspace being appurtenant to the building, then he would conclude that the airspace above the roof to the height of the chimneys is a "common part" being part of the exterior of the building.

## 4-66
### Flat
See para 4-19.

## 4-67
### Common parts
See para 4-50.

## 4-68
### Building
See para 4-17.

## 4-69

4(5) A person is a member of another's family for the purposes of this section if—

   (a) that person is the spouse or civil partner of that other person, or the two of them live together as husband and wife or as if they were civil partners, or
   (b) that person is that other person's parent, grandparent, child, grandchild, brother, sister, uncle, aunt, nephew or niece.

## 4-70

4(6) For the purposes of subsection (5)(b)—

   (a) a relationship by marriage or civil partnership shall be treated as a relationship by blood,
   (b) a relationship of the half-blood shall be treated as a relationship of the whole blood,
   (c) the stepchild of a person shall be treated as his child, and
   (d) the illegitimate child shall be treated as the legitimate child of his mother and reputed father.

## 4-71

4A. Application of right of first refusal in relation to contracts

4A(1) The provisions of this Part apply to a contract to create or transfer an estate or interest in land, whether conditional or unconditional and whether or not enforceable by specific performance, as they apply in relation to a disposal consisting of the creation or transfer of such an estate or interest.

As they so apply—

(a) references to a disposal of any description shall be construed as references to a contract to make such a disposal;
(b) references to making a disposal of any description shall be construed as references to entering into a contract to make such a disposal; and
(c) references to the transferee under the disposal shall be construed as references to the other party to the contract and include a reference to any other person to whom an estate or interest is to be granted or transferred in pursuance of the contract.

## 4-72
See para 4-46.

## 4-73

4A(2) The provisions of this Part apply to an assignment of rights under such a contract as is mentioned in subsection (1) as they apply in relation to a disposal consisting of the transfer of an estate or interest in land.

As they so apply—

(a) references to a disposal of any description shall be construed as references to an assignment of rights under a contract to make such a disposal;
(b) references to making a disposal of any description shall be construed as references to making an assignment of rights under a contract to make such a disposal;
(c) references to the landlord shall be construed as references to the assignor; and
(d) references to the transferee under the disposal shall be construed as references to the assignee of such rights.

## 4-74

Section 4A(2) states that the provisions of the 1987 Act apply to an assignment of rights under a contract and section 4A(3) states that the Act also applies to a contract to make an assignment of rights under a contract. Further, section 4A(2)(c) states that references to the landlord shall be construed as references to the assignor.

It appears that this section is intended to prevent the 1987 Act from being avoided by the following series of steps:

1. A landlord, at a time before the landlord's building is caught by the 1987 Act (for example, before it is let), enters into a contract to sell the freehold of the building to a friendly third party, Party A. The exchange of contracts is not a "relevant disposal" because the building does not fall within the Act at that time.

2. At a later date, when the building is caught by the Act, the landlord asks Party A to assign its rights under the contract (in other words, its right to take a transfer of the freehold of the building) to Party B. If it were not for section 4A(2), this assignment of rights would be outwith the Act, not being a disposal by the landlord.

3. The landlord completes the sale of the freehold of the building to Party B. Completion of the sale is not caught by the Act, because it is excluded from being a relevant disposal by section 4(2)(i) (a disposal in pursuance of a contract binding on the landlord).

The effect of the above three steps, ignoring section 4A(2), would be that the landlord had sold the freehold of the building to Party B without needing to comply with the 1987 Act, despite the fact that, at the date of the sale, the building was caught by the Act. Section 4A(2)(c), however, ensures that Party A, as the assignor of rights under the original contract, is treated as "landlord" for the purposes of the Act, and hence must comply with the obligations under section 5 in relation to the assignment of the landlord's rights.

Section 4A(4), however, states that nothing in this section affects the operation of the 1987 Act relating to options or rights of pre-emption. The grant of an option or right of pre-emption is a relevant disposal for the purposes of the Act, but completion of the sale pursuant to the exercise of an option or right of pre-emption is not a relevant disposal: section 4(2)(i). The provisions of section 4A(4), therefore, suggest that an assignment of the right to exercise an option or right of pre-emption will not be caught by the Act (not being a disposal by "the landlord"), thus opening the door to an avoidance scheme similar to that outlined above, but using a previously granted option or right of pre-emption rather than a previously exchanged contract.

## 4-75
4A(3)The provisions of this Part apply to a contract to make such an assignment as is mentioned in subsection (2) as they apply (in accordance with subsection (1)) to a contract to create or transfer an estate or interest in land.

## 4-76
4A(4)Nothing in this section affects the operation of the provisions of this Part relating to options or rights of pre-emption.

## 4-77
See para 4-73.

## 4-78
### 5. Landlord required to serve offer notice on tenants

5(1) Where the landlord proposes to make a relevant disposal affecting premises to which this Part applies, he shall serve a notice under this section (an 'offer notice') on the qualifying tenants of the flats contained in the premises (the 'constituent flats').

## 4-79
## Proposes to make a relevant disposal
Section 5(1), which requires the landlord to serve a notice where the landlord "proposes to make a relevant disposal" contrasts with section 1(1), which requires a notice to be "previously served" before the landlord may make a disposal. Section 1(1) appears to provide that, so long as notice is served before the relevant disposal is made, the landlord will have complied with its duties under the 1987 Act; whereas section 5(1) implies that the landlord has a duty to serve notice *as soon as* the landlord proposes to dispose.

Note, however, that section 10A(1) provides that the landlord will commit a criminal offence if the landlord makes the disposal without "having first complied with the requirements of section 5"—does that mean that no criminal offence is committed so long as notice is served before the disposal, or do the strict requirements of section 5(1) mean that a criminal offence may potentially be committed simply because the notice is served before the disposal is made but long after the landlord first "proposes to dispose".

Note also that, in the case of a notice served under section 5B (sale at public auction in England and Wales), the notice must be served by the landlord not less than four months and not more than six months before the date of the auction.

See para 4-47 above for the meaning of "relevant disposal".

The meaning of "proposes to make a relevant disposal" was considered by the Court of Appeal in the case of *Mainwaring* v *Henry Smith's Charity Trustees*. That case was decided in relation to the Landlord and Tenant Act 1987 as originally enacted, before the wholesale changes wrought by the Housing Act 1996, and the Court of Appeal determined that exchange of contracts was not a relevant disposal, the relevant disposal in such a case being completion of the sale (this was, of course, changed by the Housing Act 1996).

In that context, Sir Thomas Bingham MR, giving the judgment of the Court, said that section 5 notices should be served "at or within a few days of the date of ... [exchange of contracts] and possibly even sooner", on the basis of the meaning of "proposes to dispose".

This raises the question: if a landlord was previously required to serve the notices by exchange of contracts, if not sooner, in relation to a disposal that would not take place until completion (normally 28 days after exchange), when must the landlord now serve them, given that the Act now provides that exchange of contracts is itself a relevant disposal? Sir Thomas Bingham's comments are still relevant; he said that a proposal is different from an intention, and went on to say that:

> a 'proposal' ... means that a project must have moved out of 'the zone of contemplation ... into the valley of decision'.

The landlord is placed on the horns of a dilemma: on the one hand, the landlord must serve the notices as soon as the landlord proposes to dispose of the building, whereas on the other hand, the landlord needs sufficient information relating to the disposal in order to comply with the requirements of section 5, including the price, the deposit and all other principal terms of the sale contract. A mere decision to put the property on the market, therefore, is unlikely to include a decision on the principal terms of any sale contract, and it would seem that, as a minimum, heads of terms for the sale would need to be settled, at least in the landlord's mind, before notices could reasonably be served. It is highly likely, however, that some of the terms of the deal may change after heads of terms have been agreed, as part of the negotiation process, and if the landlord has already served the notices, then the landlord will need to serve fresh notices to reflect the change in the proposed terms of the disposal. If, on the other hand, the landlord waits until contracts have been agreed with a third party purchaser before serving notices, a period of at least two months will then need to elapse before contracts can be exchanged (assuming that the tenants fail to exercise their rights), and the landlord is, of course, unable to bind the third party purchaser in the meantime to go ahead on the agreed basis.

Note that the above issue will not arise in the case of a sale at a public auction in England and Wales, since section 5B(7) prescribes a specific timescale for service of the landlord's offer notice.

## 4-80
### Affecting premises
See para 4-04.

## 4-81
### Serve a notice (section 5)
Section 5 of the 1987 Act requires a landlord who is proposing to make a relevant disposal to serve a section 5 notice (an "offer notice") on the qualifying tenants. There are five different types of offer notice, depending on the type of disposal that the landlord is proposing to make, namely:

1. Section 5A: a proposed contract to create or transfer an estate or interest in land (not applicable to the grant of an option or right of pre-emption, which is covered by section 5C).

2. Section 5B: a proposed sale at a public auction to be held in England and Wales.

3. Section 5C: a proposed grant of an option or right of pre-emption.

4. Section 5D: a proposed disposal not made in pursuance of a contract, option or right of pre-emption.

5. Section 5E: a proposed disposal under sections 5A, 5B, 5C or 5D where the disposal is to be (wholly or partly) for a non-monetary consideration.

## 4-82
### Serve
Section 54(1) of the 1987 Act provides that any notice required to be served under the Act may be sent by post.

Many leases state that notices may be served in accordance with section 196 of the Law of Property Act 1925.Where this section applies (as amended by the Recorded Delivery Service Act 1962, section 1, Sch), a notice for the tenant may be served on the demised premises, and this will constitute good service even if the landlord is aware that the tenant lives elsewhere and may not therefore be aware of the notice. In such a case, it may be easiest for the landlord to arrange for somebody to leave all the notices at the building (where the individual flats do not have letter boxes, it is sufficient to leave the notices as far inside the building as the landlord's representative is able to go: see *Henry Smith's Charity Trustees* v *Kyriacou* [2001] Ch 493 at the same time, which will avoid difficulties arising where one tenant receives notice on a different date from another tenant: see para 4-92.

Note however that a notice under the 1987 Act is not a notice "required to be served under a lease", which is wording commonly used when incorporating section 196 into a lease, and if such wording is used, reliance cannot be placed on section 196 in relation to a notice under the Act, and common law will apply instead.

## 4-83
### Notice
Section 54(1) of the 1987 Act provides that any notice required to be served under the Act must be in writing.

The 1987 Act does not prescribe a form of notice, although section 54(3) permits the Secretary of State to make regulations to prescribe the form of any notice to be served under the Act (no such regulations have been made). Regulations that vary the periods specified in sections 5A(4), 5A(5), 5B(5), 5B(6), 5C(4), 5C(5), 5D(4), 5D(5) or 5E(3) can only be made under section 20(4). However, some sections of the Act do prescribe the contents of a particular type of notice. See Para 4-307 in relation to the case of *M25 Group Ltd* v *Tudor* [2003] EWCA Civ 1760, which divided the requirements set out in the Act for notice contents into those that are mandatory and those that are merely supportive (and hence non-essential). See also para 4-307 in relation to the case of *Green* v *Westleigh Properties Ltd* [2008] EWHC 1474 (QB), which applied the principles of *Mannai Investment Co Ltd* v *Eagle Star Assurance Co Ltd* [1997] AC 749 to the tenants' purchase notice.

It is good practice simply to address each notice to "The qualifying tenant" at the relevant flat, in order to avoid problems arising where a flat changes hands and the landlord has not yet been notified of the change. Care should be taken to ensure that any covering letter does not contradict the notice, and for this reason it may be preferable to avoid sending a covering letter with the notice.

The landlord may experience a number of different problems following service of section 5 notices. First, the landlord may be asked by the tenants to confirm that a requisite majority has accepted the offer, but given that the tenants may change their minds at any time before expiry of the relevant period, the position will not be clear until that period has expired. Second, information volunteered by the landlord or the landlord's agent may potentially give rise to an estoppel. From the landlord's point of view, it is preferable for the tenants to seek independent legal advice.

## 4-84
### Qualifying tenants
See para 4-37.

## 4-85
### Flat
See para 4-19.

## 4-86
### Premises
See para 4-05.

## 4-87

> 5(2) An offer notice must comply with the requirements of whichever is applicable of the following sections—
> section 5A (requirements in case of contract to be completed by conveyance, &c.),

section 5B (requirements in case of sale at auction),
section 5C (requirements in case of grant of option or right of pre-emption),
section 5D (requirements in case of conveyance not preceded by contract, &c.);

and in the case of a disposal to which section 5E applies (disposal for non-monetary consideration) shall also comply with the requirements of that section.

## 4-88
### Comply with the requirements (section 5)
Section 5 requires that the notice that corresponds to the landlord's proposed disposal be served. There is nothing in the requirements of the 1987 Act that prohibits the service of more than one type of section 5 notice nor, indeed, is there anything in the 1987 Act that deals with the practical effect of serving more than one type of notice.

There may be good reasons for a landlord wishing to serve more than one type of section 5 notice; in particular, where a property is to be sold at public auction (and a section 5B notice is therefore required), the property may in fact fail to reach its reserve and may be sold subsequently by private treaty (in which case a section 5A notice is required). Accordingly, a landlord planning to sell a property by public auction may wish to serve both a section 5B and a section 5A notice at the outset. However, there are a number of difficulties with this approach: first, the section 5A notice requires a purchase price to be stated (should this be the auction reserve price, or a figure higher or lower than that?), and second, the timing for the tenants' response and the subsequent steps to be taken is different for the two types of notice, which may lead to confusion and difficulty if the tenants wish to exercise their rights.

## 4-89
5(3) Where a landlord proposes to effect a transaction involving the disposal of an estate or interest in more than one building (whether or not involving the same estate or interest), he shall, for the purpose of complying with this section, sever the transaction so as to deal with each building separately.

## 4-90
It is particularly unfortunate, in the light of the requirement to sever any transaction dealing with more than one building, so that each building is dealt with separately, that there is no definition of "building" within the Act, as previously stated under para 4-17.

## 4-91
> 5(4) If, as a result of the offer notice being served on different tenants on different dates, the period specified in the notice as the period for accepting the offer would end on different dates, the notice shall have effect in relation to all the qualifying tenants on whom it is served as if it provided for that period to end with the latest of those dates.

## 4-92
### Different dates
It is important for the landlord to ensure that no section 5 notice is served, or deemed to be served, on a particular tenant on a date later than the date of service applicable to all the other tenants, as this will delay the start of the two-month acceptance period for all the tenants. Particular issues may arise in the case of a tenant who lives abroad, where the notice may take longer to arrive in the post or, indeed, there may be difficulties in proving the exact date on which it did arrive. Section 5(5) may assist in this situation—see para 4-94.

Many leases state that notices may be served in accordance with section 196 of the Law of Property Act 1925. Where this section applies, a notice for the tenant may be served on the demised premises, and this will constitute good service even if the landlord is aware that the tenant lives elsewhere and may not therefore be aware of the notice. For this reason, a landlord would be well advised, where section 196 applies, to serve all section 5 notices on the premises by personal delivery, as the landlord can then guarantee that the two-month acceptance period will commence, for all the tenants, on the date of personal delivery. Note however that a section 5 notice is not a notice "required to be served under a lease", which is wording commonly used when incorporating section 196 into a lease, and if such wording is used, the landlord cannot rely on section 196 in relation to section 5 notices, but must refer to common law.

## 4-93

5(5) A landlord who has not served an offer notice on all of the qualifying tenants on whom it was required to be served shall nevertheless be treated as having complied with this section—

(a) if he has served an offer notice on not less than 90% of the qualifying tenants on whom such a notice was required to be served, or
(b) where the qualifying tenants on whom it was required to be served number less than ten, if he has served such a notice on all but one of them.

## 4-94
### Treated as having complied

Note that section 5(5) of the 1987 Act, in stating that section 5 has been complied with where the landlord omits to serve notice on one qualifying tenant (where there are fewer than 10 qualifying tenants) or omits to serve notices on 10% of the qualifying tenants (where there are 10 or more qualifying tenants), effectively permits the landlord to leave certain tenants out of the service process. This means, according to section 1(1)(a), that those tenants who have not received notices, do not have rights under the 1987 Act. Potentially, therefore, a landlord could deliberately omit to serve a notice on a tenant whom the landlord believes is going to respond in a particular way and that tenant would have no right to insist on being served with a notice because that tenant has no rights under the 1987 Act *until* such notice is received.

Further, the landlord may take advantage of this provision and deliberately choose to omit serving notice on a tenant who lives abroad, in order to avoid problems in proving the date of service (or delays in service) in relation to that tenant: see para 4-92.

Note also that if the landlord has only served notices on 90% of the qualifying tenants, then a positive response is required from more than 50% of the total number of qualifying tenants (see para 4-99, "Requisite majority of qualifying tenants (section 5)"), ie more than 50% of those tenants actually served will be required, in this situation, to exercise the rights of first refusal.

## 4-95

5A. Offer notice: requirements in case of contract to be completed by conveyance, &c.

5A(1) The following requirements must be met in relation to an offer notice where the disposal consists of entering into a contract to create or transfer an estate or interest in land.

5A(2) The notice must contain particulars of the principal terms of the disposal proposed by the landlord, including in particular—

   (a) the property, and the estate or interest in that property, to which the contract relates,
   (b) the principal terms of the contract (including the deposit and consideration required).

5A(3) The notice must state that the notice constitutes an offer by the landlord to enter into a contract on those terms which may be accepted by the requisite majority of qualifying tenants of the constituent flats.

5A(4) The notice must specify a period within which that offer may be so accepted, being a period of not less than two months which is to begin with the date of service of the notice.

5A(5) The notice must specify a further period of not less than two months within which a person or persons may be nominated by the tenants under section 6.

5A(6) This section does not apply to the grant of an option or right of pre-emption (see section 5C).

## 4-96
## The following requirements must be met in relation to an offer notice (section 5A)

Where section 5A applies (that is, where the proposed disposal is an exchange of contracts), the notice must:

- contain particulars of the principal terms of the proposed disposal, including the property, the estate or interest therein proposed to be disposed of and the principal terms of the intended contract (including deposit and consideration required): section 5A(2)

- state that the notice constitutes an offer by the landlord to enter into a contract on the above terms, which may be accepted by the requisite majority of qualifying tenants: section 5A(3)

- specify a period within which the offer may be accepted, being not less than two months, starting with the date of service of the notice: section 5A(4)

- specify a further period, being not less than two months, within which the tenants may nominate a person or persons to purchase the property under section 6: section 5A(5).

Note that, although exercise of an option or of a right of pre-emption will create a contract, section 5A does not apply to this contract, as specifically stated by section 5A(6).

Note also that the two periods that must be specified must be not less than two months and so may be longer if the landlord so chooses: see *Chinnock* v *Hocaoglu* [2007] EWHC 2933 (Ch.); [2008] 2 EGLR 77, where a longer period was specified.

## 4-97
## Principal terms of the disposal/principal terms of the contract (section 5A)

The 1987 Act contains no definition of "principal terms", either in the context of the disposal itself or in terms of the contract for disposal. However, section 5A(2) makes it clear that the expression "principal terms of the disposal" includes the property, the estate or interest in that property and the principal terms of the contract, and that the phrase "principal terms of the contract" includes the deposit and consideration required.

Following delivery of the judgment in the case of *Mainwaring* v *Henry Smith's Charity Trustees*, the Court of Appeal, at a separate hearing, approved a form of offer notice which contained the following:

- Details of the property, including brief description and details of the title (the Title Number, in the case of a property registered at HM Land Registry), a plan showing the property edged in red and details of any rights to be granted or subject to which the disposal was to be made.

- The estate or interest to be disposed of, including a schedule of any leases (or other occupancy arrangements) subject to which the disposal was to be made, including the term of and annual rent reserved by each lease.

- The consideration required for the disposal, including the purchase price, deposit and any other consideration payable (such as obligations on the proposed buyer to pay rent arrears, seller's costs, etc).

- The intended completion date.

The 1987 Act contains no further detail as to what is meant by "principal terms", but it has been held, in relation to the 1987 Act before it was amended by the Housing Act 1996, that payment by an original buyer of the landlord's conveyancing costs was not a term of the disposal, in the context of an oral agreement to pay such costs (with the result that the tenants, who stepped into the shoes of the original buyer, did not have to refund these costs): see *Woodridge Ltd v Downie* [1997] 2 EGLR 193. However, the reason for this decision appears to have been based largely on the fact that the agreement to pay the landlord's costs, being oral, did not comply with the provisions of section 2 of the Law of Property (Miscellaneous Provisions) Act 1989. The landlord should be careful to avoid omitting from the offer notice something that may subsequently be held to be a principal term of the disposal, and if in doubt should err on the side of caution.

## 4-98
### Requisite majority of qualifying tenants
Section 18A of the 1987 Act effectively provides three different definitions of this phrase, depending on whether it is used in the context of section 5, section 11A or section 12A/B/C.

## 4-99
### Requisite majority of qualifying tenants (section 5)
Section 18A of the 1987 Act defines "requisite majority of qualifying tenants", for the purposes of section 5, to mean qualifying tenants having more than 50% of the available votes for flats in the premises, as at the date when the offer notice expires. Each flat occupied by a qualifying tenant is deemed to have a single vote for this purpose: section 18A(3). The Act does not deal with what would happen if a tenant of a particular flat in fact comprised more than one individual, and each individual voted differently.

Section 18A(2)(a) makes it clear that, for the purpose of calculating the available votes, a qualifying tenant is still deemed to have a vote even if the landlord failed to serve that tenant with an offer notice (see para 4-94 for the circumstances in which a landlord may comply with the Act despite failing to serve an offer notice on one or more qualifying tenants).

Note that the requisite majority of qualifying tenants must have more than 50% of the available votes. So, more than half of the number of qualifying tenants is required for this majority. This means that in the case of a building containing only two qualifying tenants, both are required to constitute a requisite majority. In the case of a building containing an even number of flats let to qualifying tenants, one more than half of those qualifying tenants is required to constitute a requisite majority, as the following table shows:

| Number of flats let to qualifying tenants in building | Number of qualifying tenants required to constitute "requisite majority" |
| --- | --- |
| 2 | 2 |
| 3 | 2 |
| 4 | 3 |
| 5 | 3 |
| 6 | 4 |
| 7 | 4 |
| 8 | 5 |
| 9 | 5 |
| 10 | 6 |

## 4-100
### Constituent flats (section 5)
This is defined in section 5(1) to mean "the flats contained in the premises". "The premises" for this purpose are the premises affected by the relevant disposal — see para 4-04.

Note that "constituent flats" is separately defined in section 11(2), where it has a slightly different meaning — see para 4-309.

## 4-101
### Offer by the landlord (section 5A)
Note that the offer contained in the landlord's offer notice is made "subject to contract" so that the tenants cannot simply accept the offer and thus create a binding contract for sale and purchase of the property by means of offer and acceptance. This is entirely consistent

with the nature of the tenants' rights, which are rights of first refusal, and hence the landlord cannot (at least until after contracts have actually been exchanged) be compelled to proceed with the disposal to the tenants, although the landlord may suffer adverse cost consequences should the landlord fail to proceed in certain circumstances—see para 4-260.

## 4-102
### 5B. Offer notice: requirements in case of sale by auction

5B(1)   The following requirements must be met in relation to an offer notice where the landlord proposes to make the disposal by means of a sale at a public auction held in England and Wales.

5B(2)   The notice must contain particulars of the principal terms of the disposal proposed by the landlord, including in particular the property to which it relates and the estate or interest in that property proposed to be disposed of.

5B(3)   The notice must state that the disposal is proposed to be made by means of a sale at a public auction.

5B(4)   The notice must state that the notice constitutes an offer by the landlord, which may be accepted by the requisite majority of qualifying tenants of the constituent flats, for the contract (if any) entered into by the landlord at the auction to have effect as if a person or persons nominated by them, and not the purchaser, had entered into it.

5B(5)   The notice must specify a period within which that offer may be so accepted, being a period of not less than two months beginning with the date of service of the notice.

5B(6)   The notice must specify a further period of not less than 28 days within which a person or persons may be nominated by the tenants under section 6.

5B(7)   The notice must be served not less than four months or more than six months before the date of the auction; and —

   (a)   the period specified in the notice as the period within which the offer may be accepted must end not less than two months before the date of the auction, and

(b) the period specified in the notice as the period within which a person may be nominated under section 6 must end not less than 28 days before the date of the auction.

5B(8) Unless the time and place of the auction and the name of the auctioneers are stated in the notice, the landlord shall, not less than 28 days before the date of the auction, serve on the requisite majority of qualifying tenants of the constituent flats a further notice stating those particulars.

## 4-103
## The following requirements must be met in relation to an offer notice (section 5B)

Where section 5B applies (that is, where the proposed disposal is by means of a sale at a public auction in England and Wales), the notice must:

- contain particulars of the principal terms of the proposed disposal, including the property, the estate or interest therein proposed to be disposed of: section 5B(2)

- state that the disposal is proposed to be by means of a sale at public auction: section 5B(3)

- state that the notice constitutes an offer by the landlord, which may be accepted by the requisite majority of qualifying tenants, for the nominated purchaser to step into the shoes of the successful bidder at the auction and acquire the property: section 5B(4)

- specify a period within which the offer may be accepted, being not less than two months, starting with the date of service of the notice: section 5B(5)

- specify a further period, being not less than 28 days, within which the tenants may nominate a person or persons to purchase the property under section 6: section 5B(6).

If the notice does not state the time and place of the auction and the name of the auctioneers, then the landlord must, at least 28 days before the auction date, serve a further notice, giving those particulars, on the requisite majority of the qualifying tenants: section 5B(8).

The proposed price is not specified in the offer notice.

Remember that there are three significant differences between section 5B (sale by public auction in England and Wales) and the other types of disposal, namely:

- In the case of section 5B, the timescale for service of the notice is not less than four months and not more than six months before the date of the auction: section 5B(7). In the case of any other type of disposal, there is no specific timetable for service of the landlord's offer notice—see para 4-79.

- In the case of section 5B, the minimum period during which the tenants may nominate a purchaser is 28 days: section 5B(6). In the case of any other type of disposal, the minimum period is two months.

- In the case of section 5B, the tenants must, in addition to responding to the offer notices and nominating a purchaser, serve an extra notice on the landlord if they wish to exercise their rights to acquire the property: see para 4-190.

## 4-104
### Sale at a public auction

There is no definition in the 1987 Act of "sale at a public auction". Although the meaning of this expression may appear to be clear, an issue may arise where a property fails to reach its reserve at the auction. In this case, a third party may subsequently approach the auctioneers and offer to buy the property at (or close to) the reserve price, whereupon the auctioneers will normally have authority from the landlord to conclude a contract for the sale of the property on the basis of the third party signing the memorandum of sale contained in the auction catalogue and providing any required deposit. The approach to the auctioneers and conclusion of the sale contract may take place:

- after the property has failed to reach its reserve, but whilst the auction is still continuing: although there is no authority on the point, it would appear to be strongly arguable that the sale has been made "at a public auction"

- after the auction has finished but whilst the auctioneers are clearing the room after the auction: again, there is no authority on the point, but there is clearly an argument that the sale has been made "at a public auction"

- one or more days after the auction has taken place, at the offices of the auctioneers (rather than at the auction venue): yet again, there is no authority on the point, but it seems difficult to argue that the sale has been made "at a public auction".

## 4-105
### Principal terms of the disposal (section 5B)
The 1987 Act contains no definition of "principal terms of the disposal". However, section 5B(2) makes it clear that the expression includes the property and the estate or interest in that property.

See also para 4-97.

## 4-106
### Requisite majority of qualifying tenants
See para 4-99.

## 4-107
### Offer by the landlord (section 5B)
Note that this is not an offer to sell the property, but an offer to allow the nominated purchaser to step into the shoes of the successful bidder at the auction. If, in fact, there is no successful bidder, because the property does not reach its reserve, then there is no contract of which the nominated purchaser can take the benefit.. For this reason, the tenants may wish to attend the auction with a view to bidding for the property up to the reserve price in order to ensure that there is a contract in existence that the nominated purchaser can take advantage of. Alternatively, the nominated purchaser may wish to bid at the auction.

## 4-108
### Serve
See para 4-82.

## 4-109
### Notice
See para 4-83.

## 4-110
### 5C. Offer notice: requirements in case of grant or option or right of pre-emption

5C(1)  The following requirements must be met in relation to an offer notice where the disposal consists of the grant of an option or right of pre-emption.

5C(2)  The notice must contain particulars of the principal terms of the disposal proposed by the landlord, including in particular—

(a) the property, and the estate or interest in that property, to which the option or right of pre-emption relates,
(b) the consideration required by the landlord for granting the option or right of pre-emption, and
(c) the principal terms on which the option or right of pre-emption would be exercisable, including the consideration payable on its exercise.

5C(3)  The notice must state that the notice constitutes an offer by the landlord to grant an option or right of pre-emption on those terms which may be accepted by the requisite majority of qualifying tenants of the constituent flats.

5C(4)  The notice must specify a period within which that offer may be so accepted, being a period of not less than two months which is to begin with the date of service of the notice.

5C(5)  The notice must specify a further period of not less than two months within which a person or persons may be nominated by the tenants under section 6.

## 4-111
## The following requirements must be met in relation to an offer notice (section 5C)

Where section 5 applies (that is, where the proposed disposal is the grant of an option or right of pre-emption), the notice must:

- contain particulars of the principal terms of the proposed disposal, including the property, the estate or interest therein proposed to be disposed of, the consideration required by the landlord for making the grant and the principal terms on which the right would be exercisable (including consideration required on exercise of the right): section 5C(2)

- state that the notice constitutes an offer by the landlord to grant an option or right of pre-emption on the above terms, which may be accepted by the requisite majority of qualifying tenants: section 5C(3)

- specify a period within which the offer may be accepted, being not less than two months, starting with the date of service of the notice: section 5C(4)

- specify a further period, being not less than two months, within which the tenants may nominate a person or persons to purchase the property under section 6: section 5C(5).

## 4-112
### Principal terms of the disposal/principal terms on which the option or right of pre-emption would be exercisable (section 5C)

The 1987 Act contains no definition of "principal terms", either in the context of the disposal itself or in respect of the terms on which the option or right of pre-emption would be exercisable. However, section 5C(2) makes it clear that the expression "principal terms of the disposal" includes the property, the estate or interest in that property, the consideration for granting the option and the principal terms on which the option/right of pre-emption is exercisable. The phrase "principal terms on which the option/right of pre-emption is exercisable" includes the consideration payable on its exercise.

See also para 4-97.

## 4-113
### Requisite majority of qualifying tenants
See para 4-99.

## 4-114
### Offer by the landlord (section 5C)
Note that the offer contained in the landlord's offer notice is made "subject to contract" so that the tenants cannot simply accept the offer and thus create a binding option agreement or agreement granting rights of pre-emption by means of offer and acceptance. This is entirely consistent with the nature of the tenants' rights, which are

rights of first refusal, and hence the landlord cannot be compelled to proceed with the disposal to the tenants, although may suffer adverse cost consequences should the landlord fail to proceed in certain circumstances — see para 4-260.

### 4-115
**5D. Offer notice: requirements in case of conveyance not preceded by contract, &c.**

5D(1) The following requirements must be met in relation to an offer notice where the disposal is not made in pursuance of a contract, option or right of pre-emption binding on the landlord.

5D(2) The notice must contain particulars of the principal terms of the disposal proposed by the landlord, including in particular—

  (a) the property to which it relates and the estate or interest in that property proposed to be disposed of, and
  (b) the consideration required by the landlord for making the disposal.

5D(3) The notice must state that the notice constitutes an offer by the landlord to dispose of the property on those terms which may be accepted by the requisite majority of qualifying tenants of the constituent flats.

5D(4) The notice must specify a period within which that offer may be so accepted, being a period of not less than two months which is to begin with the date of service of the notice.

5D(5) The notice must specify a further period of not less than two months within which a person or persons may be nominated by the tenants under section 6.

### 4-116
## The following requirements must be met in relation to an offer notice (section 5D)

Where section 5D applies (that is, where the proposed disposal constitutes completion without previously exchanging contracts or granting an option or right of pre-emption), the notice must:

- contain particulars of the principal terms of the proposed disposal, including the property, the estate or interest therein proposed to be disposed of and the consideration required: section 5D(2)

- state that the notice constitutes an offer by the landlord to dispose of the property on the above terms, which may be accepted by the requisite majority of qualifying tenants: section 5D(3)

- specify a period within which the offer may be accepted, being not less than two months, starting with the date of service of the notice: section 5D(4)

- specify a further period, being not less than two months, within which the tenants may nominate a person or persons to purchase the property under section 6: section 5D(5).

## 4-117
### Principal terms of the disposal (section 5D)
The 1987 Act contains no definition of "principal terms of the disposal". However, section 5D(2) makes it clear that the expression includes the property, the estate or interest in that property and the consideration required.

See also para 4-97.

## 4-118
### Requisite majority of qualifying tenants
See para 4-99.

## 4-119
**5E. Offer notice: disposal for non-monetary consideration**

5E(1) This section applies where, in any case to which section 5 applies, the consideration required by the landlord for making the disposal does not consist, or does not wholly consist, of money.

5E(2) The offer notice, in addition to complying with whichever is applicable of sections 5A to 5D, must state—

(a) that an election may be made under section 8C (explaining its effect), and

(b) that, accordingly, the notice also constitutes an offer by the landlord, which may be accepted by the requisite majority of

> qualifying tenants of the constituent flats, for a person or persons nominated by them to acquire the property in pursuance of sections 11 to 17.
>
> 5E(3) The notice must specify a period within which that offer may be so accepted, being a period of not less than two months which is to begin with the date of service of the notice.

## 4-120
## The offer notice ... must state

Where section 5E applies (that is, where the proposed disposal is wholly or partly for non-monetary consideration), the notice must, in addition to complying with the requirements for whichever one of sections 5A, 5B, 5C or 5D applies:

- state that an election may be made under section 8C (explaining its effect): section 5E(2)

- state that the notice constitutes an offer by the landlord, which may be accepted by the requisite majority of qualifying tenants, for the nominated purchaser to acquire the property pursuant to sections 11–17: section 5E(3)

- specify a period within which the offer may be accepted, being not less than two months, starting with the date of service of the notice: section 5E(4).

## 4-121
## Requisite majority of qualifying tenants
See para 4-99.

## 4-122
## Election
See para 4-204.

## 4-123

> **6. Acceptance of landlord's offer: general provisions**
>
> 6(1) Where a landlord has served an offer notice, he shall not during—

(a) the period specified in the notice as the period during which the offer may be accepted, or
(b) such longer period as may be agreed between him and the requisite majority of the qualifying tenants of the constituent flats,

dispose of the protected interest except to a person or persons nominated by the tenants under this section.

## 4-124
Having served offer notices under section 5 of the 1987 Act, the landlord is prohibited from proceeding with the proposed disposal, except to the nominated purchaser, until the period during which the tenants are able to accept the offer expires (a minimum of two months from service of the offer notices). The parties are able to agree to extend the period for acceptance of the offer, as referred to in section 6(1)(b). Note that, as previously stated, although the landlord is effectively embargoed from making the disposal, the landlord is equally free to choose not to make the disposal at all.

## 4-125
### Offer notice
This is a notice served by the landlord under section 5 of the 1987 Act, which is defined as an "offer notice" by section 5(1). See para 4-81 for further details.

## 4-126
### Period during which the offer may be accepted
This is either the period specified in the offer notices (which must not be shorter than a period of two months from the date of service of the offer notices), or any longer period that may be agreed between the landlord and the requisite majority of qualifying tenants. See para 4-92 where offer notices are served on different tenants on different dates.

## 4-127
### Requisite majority of qualifying tenants
See para 4-99.

## 4-128
### Constituent flats (section 5)
See para 4-100.

## 4-129
### Protected interest
Section 20(1) defines this as the estate, interest or other subject matter of an offer notice. Note that the landlord's embargo relates only to the interest of which the landlord was proposing to dispose: so if the landlord proposed to dispose of the freehold reversion, there will be no embargo in relation to the grant of a leasehold interest (although obviously such a grant may be separately caught by the 1987 Act).

## 4-130
### Person or persons nominated by the tenants
The requisite majority of qualifying tenants can nominate any person to take the disposal from the landlord. The nominated person is normally a company formed (or bought off the shelf) by the tenants for the express purpose of taking the disposal, and each of the tenants will own a share in the company. Alternatively, where there are no more than four tenants in the building, the tenants may themselves jointly be the nominated person: section 34(2) of the Trustee Act 1925 (as amended) provides that no more than four trustees may be appointed to hold land on trust for private purposes, and hence no more than four persons can jointly hold the legal estate in the building.

However, the 1987 Act lays down no requirements relating to the nominated person, except that the nominated person shall be nominated by the requisite majority. Accordingly, it would be perfectly possible for the tenants to nominate a single tenant to take the disposal alone, or indeed to nominate a person who is not even a tenant in the building. Furthermore, the Act provides no guidance as to the nomination procedure, apart from the need to serve a notice on the landlord in accordance with section 6(5) (see para 4-140).

In the case of *Mainwaring* v *Henry Smith's Charity Trustees (No 2)* [1996] EWCA Civ 657 the requisite majority of qualifying tenants purported to serve a combined acceptance and nomination notice on the landlord, nominating a company controlled by one of their number to take the disposal from the landlord. It subsequently transpired that, although the notice had originally be signed by the requisite majority, one of the tenants had in fact withdrawn his support for the nomination prior to service of the notice, despite having irrevocably agreed to join in the nomination.

The Court of Appeal held that it is not possible for individual tenants to deprive themselves of the right to withdraw their acceptance of a landlord's offer, and accordingly the irrevocable agreement was unenforceable.

This decision effectively makes it impossible for the tenants to enter into a binding agreement relating to their joint acquisition of the landlord's interest, since any of them may withdraw at any time. Yet such a binding agreement will in practice be most desirable, given that any such joint acquisition will involve the tenants incurring substantial professional costs as well as having to fund the purchase, generally by way of borrowing.

## 4-131

6(2) Where an acceptance notice is duly served on him, he shall not during the protected period (see subsection (4) below) dispose of the protected interest except to a person duly nominated for the purposes of this section by the requisite majority of qualifying tenants of the constituent flats (a 'nominated person').

## 4-132

Where the landlord receives an acceptance notice (ie a notice from the requisite majority, accepting the landlord's offer), the landlord is prevented from making the disposal, except to the nominated purchaser, during the further period allowed to the tenants under the offer notice for nomination of the nominated purchaser. This further period must not be less than two months from service of the acceptance notice, except in the case of a section 5B notice (sale by public auction in England and Wales), where the further period must not be less than 28 days from service of the acceptance notice.

## 4-133
### Acceptance notice

This is defined in section 6(3) as a notice served by the requisite majority of qualifying tenants on the landlord, accepting the offer contained in the offer notices. This acceptance is made subject to contract: section 20(2)(b).

Although section 6(3) implies that a single notice will be served, signed by each person comprising the requisite majority, in practice it is not uncommon for individual tenants to serve separate acceptance notices on their own behalf. There is nothing in the Act that specifically prohibits this, and section 6(c) of the Interpretation Act 1978 states that in any Act, unless the contrary intention appears, words in the singular include the plural and vice versa.

Individual tenants are, of course, entitled to change their minds as many times as they wish during the acceptance period, so it is dangerous for landlords to commit themselves on the basis of the first acceptance notice (or the first batch of acceptance notices) received; the Act itself envisaged this in section 9A(2) (see para 4-233), and the Court of Appeal confirmed it in *Mainwaring* v *Henry Smith's Charity Trustees (No 2)*.

The acceptance notice must be in writing and may be sent by post (section 54(1)) and must specify the names of all the persons by whom it is served and the addresses of the flats of which they are qualifying tenants (section 54(2)). The notice must also inform the landlord that the persons serving the notice accept the offer set out in the offer notices.

See the decision in *Mohammed El Naschie* v *The Pitt Place (Epsom) Ltd* [1998] NPC 83, relating to the 1987 Act before amendments were made to it by the Housing Act 1996, that an acceptance notice was void where it had been served on behalf of tenants who had not authorised this. Unfortunately, the landlord cannot force the tenants to confirm that the necessary authority has been given for the service of the acceptance notice on their behalf, and this may give rise to difficulties in certain cases.

The 1987 Act does not prescribe a form of notice, although section 54(3) permits the Secretary of State to make regulations to prescribe the form of any notice to be served under the Act (no such regulations have been made).

## 4-134
### Duly served
This is defined in section 6(3) as service within the acceptance period set out in the offer notices. See para 4-126.

See also para 4-82.

## 4-135
### Protected period
This is defined in section 6(4) as the period commencing with the date of service of the acceptance notice and ending with the end of the period for nomination of the nominated purchaser (this is the period specified in the offer notice, or any later date agreed between the landlord and the tenants). The period specified in the offer notice must be a period of not less than two months, except in the case of a section 5B notice (sale by public auction in England and Wales) where the period must be not less than 28 days: sections 5A(5), 5B(6), 5C(5) and 5D(5).

## 4-136
### Protected interest
See para 4-129.

## 4-137
### Duly nominated
Section 6(5) states that a person is "duly nominated" if that person is nominated at the same time as the acceptance notice is served or thereafter (but not after the end of the nomination period — see para 4-135).

There is no specific requirement for the nomination to be by notice, although section 8(3) deals with the landlord's obligations following service of a "notice of nomination". If a nomination notice is served, then that notice must be in writing and may be sent by post (section 54(1)) and must specify the names of all the persons by whom it is served and the addresses of the flats of which they are qualifying tenants (section 54(2)).

The 1987 Act does not prescribe a form of notice, although section 54(3) permits the Secretary of State to make regulations to prescribe the form of any notice to be served under the Act (no such regulations have been made). There is no reason why acceptance and nomination cannot be dealt with under a single form of notice.

## 4-138
### Requisite majority of qualifying tenants
See para 4-99.

## 4-139

6(3) An 'acceptance notice' means a notice served on the landlord by the requisite majority of qualifying tenants of the constituent flats informing him that the persons by whom it is served accept the offer contained in his notice.

An acceptance notice is 'duly served' if it is served within—

(a) the period specified in the offer notice as the period within which the offer may be accepted, or
(b) such longer period as may be agreed between the landlord and the requisite majority of qualifying tenants of the constituent flats.

6(4) The 'protected period' is the period beginning with the date of service of the acceptance notice and ending with—

(a) the end of the period specified in the offer notice as the period for nominating a person under this section, or
(b) such later date as may be agreed between the landlord and the requisite majority of qualifying tenants of constituent flats.

6(5) A person is 'duly nominated' for the purposes of this section if he is nominated at the same time as the acceptance notice is served or at any time after that notice is served and before the end of—

(a) the period specified in the offer notice as the period for nomination, or
(b) such longer period as may be agreed between the landlord and the requisite majority of qualifying tenants of the constituent flats.

## 4-140

Note that there is no provision for the landlord to enquire into the background to any acceptance or nomination notice. The landlord may suspect, from discussions with individual tenants, that either a requisite majority did not nominate the nominated person or a sufficient number of the requisite majority have subsequently withdrawn their support so as to reduce the number of supporting tenants below the number required for a requisite majority. However, section 9A(2) of the Act requires the nominated person to serve a notice of withdrawal on the landlord if at any time the nominated person finds that the nominated person does not have the support of the requisite majority — see para 4-233.

### 4-141
6(6) A person nominated for the purposes of this section by the requisite majority of qualifying tenants of the constituent flats may be replaced by another person so nominated if, and only if, he has (for any reason) ceased to be able to act as a nominated person.

### 4-142
Once a person has been duly nominated by the requisite majority of qualifying tenants, that person may only be replaced (by another person who has been duly nominated) where that person is no longer able to act as a nominated person. The 1987 Act gives no indication as to the circumstances in which a person would have "ceased to be able to act" as a nominated person; where the nominated person is a human being, presumably death or illness would fulfil these requirements, but normally the nominated person is a company formed (or purchased "off the shelf") for this purpose, in which case it is hard to think of circumstances that would fulfil these requirements..

### 4-143
6(7) Where two or more persons have been nominated and any of them ceases to act without being replaced, the remaining person or persons so nominated may continue to act.

### 4-144
Where more than one person was originally nominated, the remaining person(s) may continue to act as nominated person notwithstanding that one (or more) of the originally nominated person(s) has/have ceased to act and not been replaced.

### 4-145
**7. Failure to accept landlord's offer or to make nomination.**

7(1) Where a landlord has served an offer notice on the qualifying tenants of the constituent flats and—

    (a) no acceptance notice is duly served on the landlord, or

(b) no person is nominated for the purposes of section 6 during the protected period,

the landlord may, during the period of 12 months beginning with the end of that period, dispose of the protected interest to such person as he thinks fit, but subject to the following restrictions.

## 4-146
Unless the requisite majority of qualifying tenants serve an acceptance notice on the landlord within the requisite period for such acceptance and nominate a purchaser, also within the requisite period for such nomination, then the tenants lose their rights in relation to that disposal and the landlord is free to go ahead with the disposal for a period of 12 months (subject to complying with the restrictions in section 7(2) or section 7(3) (whichever is applicable)).

## 4-147
**Landlord**
See para 4-29.

## 4-148
**Offer notice**
See para 4-125.

## 4-149
**Qualifying tenants**
See para 4-37.

## 4-150
**Acceptance notice**
See para 4-133.

## 4-151
**Duly served**
See para 4-134.

## 4-152
### Person...nominated
See para 4-130.

## 4-153
### Protected period
See para 4-135.

## 4-154
### End of that period
Logic would seem to dictate that this phrase refers to the point at which the qualifying tenants lose their rights in relation to the relevant disposal. In other words, if they fail to accept the offer within the acceptance period, then "the end of that period" means the end of the acceptance period, whereas if they duly accept the offer but then fail to make the nomination within the protected period, then "the end of that period" means the end of the protected period.

However, the strict wording of the Act implies that "the end of that period" will always be the end of the protected period. It would seem illogical, however, to prevent the landlord from proceeding with the disposal at the end of the acceptance period, the tenants having failed to exercise their rights, and forcing the landlord to wait at least a further two months (or, in the case of section 5B, 28 days) after the end of the acceptance period.

## 4-155
### Protected interest
See para 4-129.

## 4-156
### Following restrictions
These are the restrictions referred to in section 7(2) (in the case of a sale by auction) or in section 7(3) (in the case of any other type of disposal).

## 4-157
7(2)   Where the offer notice was one to which section 5B applied (sale by auction), the restrictions are —

> > (a) that the disposal is made by means of a sale at a public auction, and
> > (b) that the other terms correspond to those specified in the offer notice.
>
> 7(3) In any other case the restrictions are—
>
> > (a) that the deposit and consideration required are not less than those specified in the offer notice, and
> > (b) that the other terms correspond to those specified in the offer notice.
>
> 7(4) The entitlement of a landlord, by virtue of this section or any other corresponding provision of this Part, to dispose of the protected interest during a specified period of 12 months extends only to a disposal of that interest, and accordingly the requirements of section 1(1) must be satisfied with respect to any other disposal by him during that period of 12 months (unless the disposal is not a relevant disposal affecting any premises to which at the time of the disposal this Part applies).

## 4-158

Although the landlord is free to proceed with the disposal after the tenants have lost their rights in relation to it, the 1987 Act protects the tenants by preventing the landlord from disposing on different terms or in a different way. So, where the offer notice related to a public auction sale in England and Wales, the landlord's subsequent disposal must also be by way of sale at a public auction (although not necessarily in England and Wales), where the offer notice related to any other disposal, the deposit and consideration required must not be less than those stated in the offer notices (although they may be greater) and, in every case, the other terms of the disposal must correspond to those in the offer notice. Further, the landlord's freedom to dispose relates only to the interest originally proposed to be disposed of by the landlord, as referred to in the offer notices.

## 4-159

> **8. Landlord's obligations in case of acceptance and nomination.**
>
> 8(1) This section applies where a landlord serves an offer notice on the qualifying tenants of the constituent flats and —

(a) an acceptance notice is duly served on him, and
(b) a person is duly nominated for the purposes of section 6,

by the requisite majority of qualifying tenants of the constituent flats.

## 4-160
This deals with the landlord's obligations where the tenants have protected their position by serving an acceptance notice and nominating a purchaser, namely not to proceed with the proposed disposal (except to the nominated purchaser) and within one month of the date of service of the nomination notice, either to proceed with the disposal to the nominated person (complying with section 8A, section 8B, section 8C or section 8D, whichever is applicable) or not to proceed with the disposal.

## 4-161
### Landlord
See para 4-29.

## 4-162
### Offer notice
See para 4-125.

## 4-163
### Qualifying tenants
See para 4-37.

## 4-164
### Acceptance notice
See para 4-133.

## 4-165
### Duly served
See para 4-134.

## 4-166
### Duly nominated
See para 4-137.

## 4-167
### Requisite majority of qualifying tenants
See para 4-99.

## 4-168
8(2) Subject to the following provisions of this Part, the landlord shall not dispose of the protected interest except to the nominated person.

## 4-169
### Protected interest
See para 4-129.

## 4-170
### Nominated person
See para 4-130.

## 4-171
8(3) The landlord shall, within the period of one month beginning with the date of service of notice of nomination, either—

(a) serve notice on the nominated person indicating an intention no longer to proceed with the disposal of the protected interest, or
(b) be obliged to proceed in accordance with the following provisions of this Part.

## 4-172
The landlord's notice served under (a) above is defined as "a notice of withdrawal" under section 8(4).

## 4-173
8(4) A notice under subsection (3)(a) is a notice of withdrawal for the purposes of section 9B(2) to (4) (consequences of notice of withdrawal by landlord).

## 4-174
This section defines the notice served by the landlord under section 8(3)(a), stating the landlord's intention not to proceed with the disposal, as "a notice of withdrawal".

## 4-175
8(5)   Nothing in this section shall be taken as prejudicing the application of the provisions of this Part to any further offer notice served by the landlord on the qualifying tenants of the constituent flats.

## 4-176
**8A.   Landlord's obligation: general provisions.**

8A(1)   This section applies where the landlord is obliged to proceed and the offer notice was not one to which section 5B applied (sale by auction).

## 4-177
Section 8A only applies where (i) the requisite majority of qualifying tenants have served an acceptance notice, and (ii) a purchaser has been nominated, and (iii) the offer notice did not relate to a proposed sale by public auction in England and Wales.

Note that section 8A applies to every type of proposed disposal, except for a sale by public auction in England and Wales. As a result, an exchange of contracts is envisaged in every case, not only where the proposed disposal was itself an exchange of contracts, but also where the proposed disposal involved the grant of an option or right of pre-emption or the completion of a conveyance (not preceded by a contract for sale). There is no particular problem in the requirement for a contract to precede a disposal by way of conveyance, but it seems illogical, in the case of the grant of an option, to require a contract for the grant of the option preceding the actual grant itself when an option is normally granted by way of an option agreement (a species of contract).

## 4-178
### Obliged to proceed
It is misleading to refer to the landlord being "obliged to proceed"; the landlord is never obliged to proceed with the disposal, but may suffer

adverse cost consequences if the landlord withdraws from the disposal to the nominated purchaser at a late stage (and, having withdrawn from the disposal to the nominated purchaser, is not free to dispose of the relevant interest to a third party): see the comments of Laddie J in the case of *Michaels* v *Taylor Woodrow Developments Ltd* [2000] PLSCS 101 at para 1–3:

> Although it has been said that [the 1987 Act] gives rise to a right of first refusal, the tenants' rights are somewhat more limited than that...the tenants acquire by the Section 5 notice an ability to be considered as potential purchasers of the interest which the landlord wishes to dispose of, but no entitlement to force the landlord to sell to them.

### 4-179

    8A(2) The landlord shall, within the period of one month beginning with the date of service of the notice of nomination, send to the nominated person a form of contract for the acquisition of the protected interest on the terms specified in the landlord's offer notice.

    8A(3) If he fails to do so, the following provisions of this Part apply as if he had given notice under section 9B (notice of withdrawal by landlord) at the end of that period.

### 4-180

Unless the landlord sends a form of contract (complying with the terms set out in the offer notice) to the nominated person within one month of the date of service of the notice of nomination, the landlord is deemed to have served a notice of withdrawal under section 9B at the end of that one month period.

### 4-181

    8A(4) If the landlord complies with subsection (2), the nominated person shall, within the period of two months beginning with the date on which it is sent or such longer period beginning with that date as may be agreed between the landlord and that person, either—

        (a) serve notice on the landlord indicating an intention no longer to proceed with the acquisition of the protected interest, or

*Section by Section*

    (b)    offer an exchange of contracts, that is to say, sign the contract and send it to the landlord, together with the requisite deposit.

In this subsection 'the requisite deposit' means a deposit of an amount determined by or under the contract or an amount equal to 10 per cent of the consideration, whichever is the less.

8A(5)    If the nominated person—

    (a)    serves notice in pursuance of paragraph (a) of subsection (4), or
    (b)    fails to offer an exchange of contracts within the period specified in that subsection,

the following provisions of this Part apply as if he had given notice under section 9A (withdrawal by nominated person) at the same time as that notice or, as the case may be, at the end of that period.

## 4-182

The nominated purchaser has two months from the date of receipt of the contract from the landlord (or any longer period that may be agreed between the parties) either (i) to notify the landlord that the nominated purchaser no longer intends to proceed with the acquisition or (ii) to send to the landlord the signed contract together with the requisite deposit (defined in section 8(4) as the deposit provided for in the contract or 10% of the consideration, whichever is the lower amount). If the nominated purchaser notifies the landlord under (i) above, or fails to send through the signed contract and deposit under (ii) above, the nominated purchaser is deemed to have served a notice of withdrawal under section 9A, either at the date when the nominated purchaser notified the landlord under (i) above or at the end of the period for sending through the contract and deposit, whichever is applicable.

## 4-183

8A(6)    If the nominated person offers an exchange of contracts within the period specified in subsection (4), but the landlord fails to complete the exchange within the period of seven days beginning with the day on which he received that person's contract, the following provisions of this Part apply as if the landlord had given notice under section 9B (withdrawal by landlord) at the end of that period.

## 4-184

As previously stated, the landlord can never be compelled to proceed with the disposal to the nominated purchaser, unless and until contracts for that disposal have been exchanged. Even if the nominated purchaser sends through the signed contract and deposit, the landlord may still choose not to exchange contracts (although there may be adverse cost consequences for the landlord and the landlord will not then be free to make that disposal).

## 4-185

    **8B.    Landlord's obligation: election in case of sale at auction.**

    8B(1)    This section applies where the landlord is obliged to proceed and the offer notice was one to which section 5B applied (sale by auction).

## 4-186

Section 8B only applies where (i) the requisite majority of qualifying tenants have served an acceptance notice, and (ii) a purchaser has been nominated, and (iii) the offer notice related to a proposed sale by public auction in England and Wales. Note that in the case of the sale being at auction (but not in the case of any other disposal), there is an additional hurdle for the tenants to clear if they are to exercise their rights—see para 4-190—otherwise the landlord is not obliged to proceed under section 8B.

## 4-187
### Obliged to proceed
See para 4-178.

## 4-188

    8B(2)    The nominated person may, by notice served on the landlord not less than 28 days before the date of the auction, elect that the provisions of this section shall apply.

## 4-189
### Nominated person
See para 4-130.

## 4-190
### May
The use of the word "may" is extremely misleading; in fact, failure to serve such a notice will result in the tenants/nominated person losing their rights under the 1987 Act in relation to the proposed disposal — see section 8B(6).

## 4-191
### Notice
See para 4-83.

## 4-192
### Served
See para 4-82.

## 4-193
> 8B(3) If a contract for the disposal is entered into at the auction, the landlord shall, within the period of seven days beginning with the date of the auction, send a copy of the contract to the nominated person.

## 4-194
### Copy of the contract
The contract is formed on the fall of the auctioneer's hammer, and is therefore unwritten. However, it is usual for the successful bidder to sign the memorandum of sale within the auction catalogue, and presumably it is a copy of this memorandum (together with a copy of the conditions of sale referred to therein) that the landlord is expected to send to the nominated person.

## 4-195
> 8B(4) If, within the period of 28 days beginning with the date on which such a copy is so sent, the nominated person—
>
> (a) serves notice on the landlord accepting the terms of the contract, and
> (b) fulfils any conditions falling to be fulfilled by the purchaser on entering into the contract,

the contract shall have effect as if the nominated person, and not the purchaser, had entered into the contract.

## 4-196
The nominated person may step into the shoes of the successful bidder at the auction and become the contracting purchaser under the auction contract by (i) serving notice on the landlord accepting the terms of the contract and (ii) fulfilling any conditions that the purchaser must fulfil under the contract (eg the payment of a deposit). Both steps must be taken within the period of 28 days from the date on which a copy of the auction contract was sent to the nominated person.

See para 4-199 in relation to the effect of section 8B(4) on any time limit in the contract.

See para 4-201 in relation to the consequences of failure by the nominated person to take the steps referred to in section 8B(4).

## 4-197
### The contract shall have effect... (section 8B(4))
The wording of section 8B(4) implies that the legal effect of the contract between the landlord and the successful bidder is altered so that for all purposes, the contract now takes effect between landlord and tenants rather than landlord and bidder. This begs the question: if the successful bidder no longer has the benefit of the contract, does the successful bidder have any remedy against the landlord if the bidder was unaware of the tenants' rights? This will depend on any representations made by the landlord to the bidder, and the bidder may well be able to bring a claim for misrepresentation against the landlord if, for example, the landlord confirmed in replies to enquiries before contract before the auction that the property did not fall within the 1987 Act. As the bidder no longer has the benefit of the contract, it would appear that the bidder cannot bring a claim under that contract, for example in relation to any provisions in the auction sale conditions.

## 4-198
8B(5)  Unless otherwise agreed, any time limit in the contract as it has effect by virtue of subsection (4) shall start to run again on the service of notice under that subsection; and nothing in the contract as it has effect by virtue of a notice under this section shall require the nominated person to complete the purchase before the end of the period of 28 days beginning with the day on which he is deemed to have entered into the contract.

## 4-199

Having stepped into the shoes of the successful bidder at the auction under section 8B(4), the nominated person is obliged to comply with the obligations under the auction contract. However, up to 35 days (seven days under section 8B(3) and 28 days under section 8B(4)) from the date of the auction may have elapsed before the nominated person is substituted for the successful bidder. For this reason, the nominated person cannot be expected to comply with the same time limits, or to complete the purchase on the same completion date, as the successful bidder.

This section therefore provides (i) that any time limit in the auction contract will start to run again on the service of the nominated person's notice under section 8B(4); and (ii) that the completion date cannot occur less than 28 days after "the day on which [the nominated person] is deemed to have entered into the contract" (see para 4-200).

## 4-200
## The day on which [the nominated person] is deemed to have entered into the contract

Section 8B(4) states that the contract shall have effect as if the nominated person had been the contracting party (rather than the successful bidder), provided that the nominated person serves notice and complies with any conditions within 28 days of receiving a copy of the auction contract. This suggests that the nominated person is deemed to have entered into the contract on the date on which the nominated person serves notice and complies with any conditions.

Where notice is served on one day and the conditions are complied with on a different day, common sense would suggest that the contract is deemed to have been entered into on the later of the two dates. On the other hand, section 8B(5) provides for time limits in the contract to start to run again on the date on which the nominated person serves notice under section 8B(4). There is no reference to compliance with any conditions, nor to the fact that this may take place after service of notice.

Furthermore, section 8B(4) states that the auction contract shall have effect as if the nominated person had entered into it. The auction contract was created at the auction on the fall of the hammer, and accordingly would have been dated with the date of the auction. Could this mean that the nominated person is deemed to have entered into the auction contract on the auction date? If so, the provision in section 8B(5) that the nominated person cannot be required to complete less than 28 days after the nominated person is deemed to have entered into the auction contract may give rise to difficulties. If the 28-day period is calculated from the auction date, then the nominated person may actually be obliged to complete the purchase at a date before serving notice under section 8B(4)(a). This is illogical, and presumably Parliament's intention was that the nominated person is effectively deemed to have entered into the original auction contract but with a new contract date.

### 4-201

8B(6) If the nominated person—

(a) does not serve notice on the landlord under subsection (2) by the time mentioned in that subsection, or
(b) does not satisfy the requirements of subsection (4) within the period mentioned in that subsection,

the following provisions of this Part apply as if he had given notice under section 9A (withdrawal by nominated person) at the end of that period.

If the nominated person fails to serve notice under section 8B(2) or fails to satisfy the requirements of section 8B(4), the nominated person is deemed to have served notice of withdrawal on the landlord under section 9A.

### 4-202
### At the end of that period

Section 8B(6)(a) refers to "the time mentioned" in section 8B(2) and section 8B(6)(b) refers to "the period mentioned" in section 8B(4). That suggests that the nominated person is deemed to have withdrawn at the end of the period in (b). However, the period in (b) arises from the provisions of section 8B(4), which does not apply unless the nominated person has served notice under section 8B(2). As a result, "at the end of that period" must logically mean either by the end of the time mentioned in section 8B(2) or at the end of the period mentioned in section 8B(4), whichever is applicable.

## 4-203
**8C. Election in case of disposal for non-monetary consideration.**

8C(1) This section applies where an acceptance notice is duly served on the landlord indicating an intention to accept the offer referred to in section 5E (offer notice: disposal for non-monetary consideration).

8C(2) The requisite majority of qualifying tenants of the constituent flats may, by notice served on the landlord within—

(a) the period specified in the offer notice for nominating a person or persons for the purposes of section 6, or
(b) such longer period as may be agreed between the landlord and the requisite majority of qualifying tenants of the constituent flats,

elect that the following provisions shall apply.

8C(3) Where such an election is made and the landlord disposes of the protected interest on terms corresponding to those specified in his offer notice in accordance with section 5A, 5B, 5C or 5D, sections 11 to 17 shall have effect as if—

(a) no notice under section 5 had been served;
(b) in section 11A(3)(period for serving notice requiring information, &c.), the reference to four months were a reference to 28 days; and
(c) in section 12A(2) and 12B(3)(period for exercise of tenants' rights against purchaser) each reference to six months were a reference to two months.

8C(4) For the purposes of sections 11 to 17 as they have effect by virtue of subsection (3) so much of the consideration for the original disposal as did not consist of money shall be treated as such amount in money as was equivalent to its value in the hands of the landlord.

The landlord or the nominated person may apply to have that amount determined by a leasehold valuation tribunal.

## 4-204
This is an anti-avoidance provision. Without section 8C, the landlord could agree a disposal of premises to a third party in consideration of

some valuable item that the nominated person could not provide, such as a specified property owned by the third party or a specified work of art. The disposal would still be caught by the 1987 Act, but the tenants would in practice be unable to exercise their rights. Section 8C, however, provides for the disposal to proceed to a third party, and then for the nominated person to acquire the premises from that third party in return for the value of the non-monetary consideration in the hands of the landlord (plus any monetary consideration paid).

Where the landlord's proposed disposal is to be made, wholly or partly, for non-monetary consideration, the landlord must comply with section 5E (see para 4-120). This is in addition to compliance with any section 5A, 5B, 5C or 5D obligations. If the requisite majority of qualifying tenants wish to exercise their rights under section 5E and make the election offered to them by the landlord, they must serve both an "acceptance notice" (see para 4-133) under section 8C(1) and also a further notice under section 8C(2). The notice under section 8C(2) must be served within the "protected period" (see para 4-135). In addition, of course, the tenants must serve an acceptance notice in relation to the offer made under section 5A or 5B or 5C or 5D as the case may be, as well as nominating a person under section 6(5); there does not appear to be any reason why a single acceptance notice should not be served.

Where the tenants exercise their rights as stated above, the landlord will proceed to carry out the disposal to a third party. The nominated person will then be able to exercise its rights against the third party under sections 11 to 17 of the 1987 Act.

Note that it may in fact be possible for the landlord to require a monetary consideration for a proposed disposal that the nominated person is unable to provide, for example, moneys paid from a specific numbered bank account. The 1987 Act does not contain anti-avoidance provisions dealing with this problem.

Section 8C(3) implies that the landlord is free to make the disposal to the third party in order that the tenants can exercise their rights against that third party, but such a disposal will, strictly speaking, be in breach of the Act. It may be that the possibility of the landlord being guilty of a criminal offence in such circumstances is avoided because of the words "without reasonable excuse" in section 10A(1) (see para 4-283).

## 4-205
## Requisite majority of qualifying tenants
See para 4-99.

## 4-206
## Notice
See para 4-83.

## 4-207
## Served
See para 4-82.

## 4-208
## Determined by a Leasehold Valuation Tribunal
Note that only the landlord or the nominated person may apply to a Tribunal to have the value of the non-monetary consideration in the hands of the landlord determined. Despite the fact that it is the third party purchaser who will receive the consideration from the nominated person, the third party purchaser has no right to apply to the Tribunal.

## 4-209
**8D. Disposal in pursuance of option or right of pre-emption.**

8D(1) Where—

(a) the original disposal was the grant of an option or right of pre-emption, and
(b) in pursuance of the option or right, the landlord makes another disposal affecting the premises ( 'the later disposal') before the end of the period specified in subsection (2),

sections 11 to 17 shall have effect as if the later disposal, and not the original disposal, were the relevant disposal.

## 4-210
Section 8D deals with the situation where the landlord, in breach of the 1987 Act, makes a disposal consisting of the grant of an option or a right of pre-emption, and before the tenants have been able to exercise

their rights in relation to that disposal, the landlord makes a further disposal (presumably completion of a transfer or grant of lease pursuant to the exercise of the option or right of pre-emption). In such a case, there is little point in the tenants exercising rights in relation to the first disposal, since this would entitle them merely to the benefit of an option or pre-emption that has already been exercised and has therefore ceased to exist. The section therefore provides that the tenants' rights apply to the subsequent disposal, thus enabling the tenants to get their hands on the premises or the lease granted pursuant to the exercise of the option or right of pre-emption. The timing for the tenants to exercise their rights in relation to the subsequent disposal is the same as that in relation to any disposal made in breach of the 1987 Act—see para 4-314.

## 4-211

8D(2) The period referred to in subsection (1)(b) is the period of four months beginning with the date by which—

(a) notices under section 3A of the Landlord and Tenant Act 1985 (duty of new landlord to inform tenants of rights) relating to the original disposal, or

(b) where that section does not apply, documents of any other description—

(i) indicating that the original disposal has taken place, and

(ii) alerting the tenants to the existence of their rights under this Part and the time within which any such rights must be exercised,

have been served on the requisite majority of qualifying tenants of the constituent flats.

## 4-212

**8E. Covenant, &c affecting landlord's power to dispose.**

8E(1) Where the landlord is obliged to proceed but is precluded by a covenant, condition or other obligation from disposing of the protected interest to the nominated person unless the consent of some other person is obtained—

(a) he shall use his best endeavours to secure that the consent of that person to that disposal is given, and

(b) if it appears to him that that person is obliged not to withhold his consent unreasonably but has nevertheless so withheld it, he shall institute proceedings for a declaration to that effect.

## 4-213

Section 8E(1) deals with the situation where the consent of a third party (such as a superior landlord) is required before the landlord may dispose of the premises to the nominated person. It obliges the landlord to use best endeavours to obtain that consent and to seek declaratory relief where the landlord believes that consent is being unreasonably withheld despite the third party being obliged not unreasonably to withhold consent.

If the landlord's application for declaratory relief is unsuccessful, the landlord may notify the tenant under section 8E(3) and the landlord is then free for a period of 12 months from the date of service of that notice to make the proposed disposal to a third party subject to the restrictions in section 8E(4) (in the case of a sale at public auction) or section 8E(5) (in any other case). The landlord may also be entitled to recover costs from the tenants/nominated person under section 8E(6).

The obligations to use best endeavours is an onerous one. In the case of *Sheffield District Railway Company* v *Great Central Railway Company* [1911] 27 TLR 451, Lawrence J said:

> 'Best endeavours' means what the words say: they do not mean second best endeavours....they do not mean that the limits of reason must be overstepped...but short of these qualifications the words mean that [the person obliged to use best endeavours] must, broadly speaking, leave no stone unturned...

## 4-214

8E(2) Subsection (1) ceases to apply if a notice of withdrawal is served under section 9A or 9B (withdrawal of either party from transaction) or if notice is served under section 10 (lapse of landlord's offer: premises ceasing to be premises to which this Part applies).

## 4-215

The landlord's obligations under section 8E(1) cease to apply where either party withdraws from the transaction under section 9A or section 9B, or where the premises cease to fall within the 1987 Act under section 10.

## 4-216

8E(3) Where the landlord has discharged any duty imposed on him by subsection (1) but any such consent as is there mentioned has been withheld, and no such declaration as is there mentioned has been made, the landlord may serve a notice on the nominated person stating that to be the case.

When such a notice has been served, the landlord may, during the period of 12 months beginning with the date of service of the notice, dispose of the protected interest to such person as he thinks fit, but subject to the following restrictions.

## 4-217
See para 4-213.

## 4-218

8E(4) Where the offer notice was one to which section 5B applied (sale by auction), the restrictions are—

(a) that the disposal is made by means of a sale at a public auction, and
(b) that the other terms correspond to those specified in the offer notice.

8E(5) In any other case the restrictions are—

(a) that the deposit and consideration required are not less than those specified in the offer notice or, if higher, those agreed between the landlord and the nominated person (subject to contract), and
(b) that the other terms correspond to those specified in the offer notice.

## 4-219
## Or, if higher, those agreed between the landlord and the nominated person (subject to contract)
See para 4-239.

## 4-220

8E(6) Where notice is given under subsection (3), the landlord may recover from the nominated party and the qualifying tenants who served the acceptance notice any costs reasonably incurred by him in connection with the disposal between the end of the first four weeks of the nomination period and the time when that notice is served by him.

Any such liability of the nominated person and those tenants is a joint and several liability.

## 4-221
### Acceptance notice
See para 4-133.

## 4-222
### In connection with the disposal
See para 4-243.

## 4-223
### Nomination period
See para 4-135.

## 4-224
### Joint and several liability
See para 4-245.

## 4-225
**9A. Notice of withdrawal by nominated person.**

9A(1) Where the landlord is obliged to proceed, the nominated person may serve notice on the landlord (a 'notice of withdrawal') indicating his intention no longer to proceed with the acquisition of the protected interest.

## 4-226
Section 9A(1) gives the nominated person the right (but does not impose any obligation) to serve notice on the landlord withdrawing from the transaction (defined as a "notice of withdrawal").

## 4-227
### Obliged to proceed
See para 4-178.

## 4-228
### Nominated person
See para 4-130.

## 4-229
### Serve
See para 4-82.

## 4-230
### Notice
See para 4-83.

## 4-231
### Protected interest
See para 4-129.

## 4-232
> 9A(2) If at any time the nominated person becomes aware that the number of the qualifying tenants of the constituent flats desiring to proceed with the acquisition of the protected interest is less than the requisite majority of qualifying tenants of those flats, he shall forthwith serve a notice of withdrawal.

## 4-233
Where section 9A(1) is optional, section 9A(2) is mandatory. It obliges the nominated person to serve a notice of withdrawal as soon as the nominated person becomes aware that the number of supporting tenants has fallen below the requisite majority. See also para 4-130 in relation to the tenants' right to withdraw their support.

## 4-234
### Requisite majority of qualifying tenants
See para 4-99.

## 4-235
## Notice of withdrawal (nominated person)
As stated in para 4-226, section 9A(1) defines the notice served on the landlord by the nominated person as a "notice of withdrawal".

## 4-236
9A(3)   Where notice of withdrawal is given by the nominated person under this section, the landlord may, during the period of 12 months beginning with the date of service of the notice, dispose of the protected interest to such person as he thinks fit, but subject to the following restrictions.

## 4-237
Following receipt of the nominated person's notice of withdrawal, the landlord is free, for a period of 12 months from the date of service of the notice of withdrawal, to proceed with the proposed disposal to a third party, subject to the restrictions in section 9A(4) (in the case of a sale at public auction) or section 9A(5) (in any other case).

## 4-238
9A(4)   Where the offer notice was one to which section 5B applied (sale by auction), the restrictions are—

   (a) that the disposal is made by means of a sale at a public auction, and
   (b) that the other terms correspond to those specified in the offer notice.

9A(5)   In any other case the restrictions are—

   (a) that the deposit and consideration required are not less than those specified in the offer notice or, if higher, those agreed between the landlord and the nominated person (subject to contract), and
   (b) that the other terms correspond to those specified in the offer notice.

## 4-239
### Or, if higher, those agreed between the landlord and the nominated person (subject to contract)

When the landlord is permitted to make the proposed disposal to a third party, the landlord is normally placed under an obligation not to make that disposal at a price lower than that specified in the offer notice. This is reasonable since otherwise, the landlord could specify an unrealistically high price in the offer notice and, when the tenants decide not to exercise their rights at such a high price, could then reduce that price to market level when dealing with a third party purchaser.

These words imply that the landlord and the nominated person may have agreed a higher price, subject to contract, than that contained in the offer notice. It is hard to see why the nominated person should agree to a higher price, because if the value of the property increases after the offer notice is served, the landlord must either proceed with the disposal to the nominated person at the original price or the landlord will be unable to make the disposal for a period of 12 months (see para 4-257).

## 4-240

> 9A(6) If notice of withdrawal is served under this section before the end of the first four weeks of the nomination period specified in the offer notice, the nominated person and the qualifying tenants who served the acceptance notice are not liable for any costs incurred by the landlord in connection with the disposal.
>
> 9A(7) If notice of withdrawal is served under this section after the end of those four weeks, the landlord may recover from the nominated person and the qualifying tenants who served the acceptance notice any costs reasonably incurred by him in connection with any disposal between the end of those four weeks and the time when the notice of withdrawal was served on him.
>
> Any such liability of the nominated person and those tenants is a joint and several liability.

## 4-241

Provided that the nominated person serves notice of withdrawal within four weeks of the start of the nomination period, there is no possibility of either the nominated person or the qualifying tenants having to pay the landlord's costs. Any later withdrawal, however, may bring a joint and several liability on the nominated person and those qualifying tenants who served the acceptance notice for the landlord's costs.

## 4-242
### Acceptance notice
See para 4-133.

## 4-243
### In connection with the disposal
Although it would seem to be logical for the landlord only to be able to recover costs insofar as they relate to the proposed disposal to the nominated person, these words imply that the landlord can in fact recover costs, provided that they are reasonably incurred, in connection with the proposed disposal. Conceivably this could include costs incurred in dealing with third parties, particularly in the case of a sale at auction. It seems unlikely that this would have been Parliament's intention.

## 4-244
### Nomination period
See para 4-135.

## 4-245
### Joint and several liability
The liability of the tenants who served an acceptance notice and the liability of the nominated person is joint and several. This means that the landlord can pursue any one or more of them for costs, although the landlord cannot recover costs more than once.

## 4-246
9A(8) This section does not apply after a binding contract for the disposal of the protected interest—

(a) has been entered into by the landlord and the nominated person, or
(b) has otherwise come into existence between the landlord and the nominated person by virtue of any provision of this Part.

## 4-247

Once contracts for the proposed disposal have been exchanged between the landlord and the nominated person, the law of contract (including remedies relating to failure to complete, such as specific performance, damages, etc) applies to that contract and the 1987 Act (and its applicability or otherwise) becomes irrelevant (except where the landlord rescinds the contract under section 10(5)).

## 4-248
### Has otherwise come into existence

Section 9A(8)(a) deals with the situation where contracts have been exchanged between the landlord and the nominated person for the disposal of the protected interest. Section 9A(8)(b) deals with the situation where a contract has come into existence between the landlord and the nominated person without contracts being formally exchanged. The only situation where (b) would apply is where the nominated person has exercised rights under section 8B(4).

## 4-249

    **9B.**    **Notice of withdrawal by landlord.**

    9B(1)    Where the landlord is obliged to proceed, he may serve notice on the nominated person (a 'notice of withdrawal') indicating his intention no longer to proceed with the disposal of the protected interest.

## 4-250

Section 9B(1) gives the landlord the right (but does not impose any obligation) to serve notice on the nominated person withdrawing from the transaction (defined as a "notice of withdrawal"). Unlike in the case of the nominated person (see para 4-235), the landlord is never *obliged* to serve a notice of withdrawal—see in particular para 4-267.

## 4-251
### Obliged to proceed
See para 4-178.

## 4-252
**Serve**
See para 4-82.

## 4-253
**Notice**
See para 4-83.

## 4-254
**Nominated person**
See para 4-130.

## 4-255
**Protected interest**
See para 4-129.

## 4-256
9B(2)    Where a notice of withdrawal is given by the landlord, he is not entitled to dispose of the protected interest during the period of 12 months beginning with the date of service of the notice.

## 4-257
Following service of the landlord's notice of withdrawal, the landlord is prevented, for a period of 12 months from the date of service of the notice of withdrawal, from disposing of the protected interest. Note that the landlord is not, however, prevented from disposing of a similar interest in the same premises during that period—for example, where the protected interest was the freehold, the landlord may grant a 999-year lease of the same premises (subject, of course, to complying with the requirements of the 1987 Act).

## 4-258
**Notice of withdrawal (landlord)**
As stated in para 4-250, section 9B(1) defines the notice served by the landlord on the nominated person as a "notice of withdrawal".

## 4-259

9B(3)  If a notice of withdrawal is served before the end of the first four weeks of the nomination period specified in the offer notice, the landlord is not liable for any costs incurred in connection with the disposal by the nominated person and the qualifying tenants who served the acceptance notice.

9B(4)  If a notice of withdrawal is served after the end of those four weeks, the nominated person and the qualifying tenants who served the acceptance notice may recover from the landlord any costs reasonably incurred by them in connection with the disposal between the end of those four weeks and the time when the notice of withdrawal was served.

## 4-260

Provided that the landlord serves notice of withdrawal within four weeks of the start of the nomination period, there is no possibility of the landlord having to pay the costs of either the nominated person or the qualifying tenants. Any later withdrawal, however, may bring a liability on the landlord for the costs reasonably incurred in connection with the disposal by the nominated person and those qualifying tenants who served the acceptance notice.

## 4-261
### Acceptance notice
See para 4-133.

## 4-262
### Nomination period
See para 4-135.

## 4-263

9B(5)  This section does not apply after a binding contract for the disposal of the protected interest—

(a) has been entered into by the landlord and the nominated person, or
(b) has otherwise come into existence between the landlord and the nominated person by virtue of any provision of this Part.

## 4-264

Once contracts for the proposed disposal have been exchanged between the landlord and the nominated person, the law of contract (including remedies relating to failure to complete, such as specific performance, damages, etc) applies to that contract and the 1987 Act (and its applicability or otherwise) becomes irrelevant (except where the landlord rescinds the contract under section 10(5)).

## 4-265
### Has otherwise come into existence
See para 4-248.

## 4-266
### 10. Lapse of landlord's offer.

10(1) If after a landlord has served an offer notice the premises concerned cease to be premises to which this Part applies, the landlord may serve a notice on the qualifying tenants of the constituent flats stating—

(a) that the premises have ceased to be premises to which this Part applies, and
(b) that the offer notice, and anything done in pursuance of it, is to be treated as not having been served or done;

and on the service of such a notice the provisions of this Part cease to have effect in relation to that disposal.

## 4-267

As previously stated, the landlord is never obliged to serve a notice withdrawing from the transaction. Section 10(1) makes it clear that this principle still applies even where the 1987 Act ceases to apply to the premises after the service of the offer notice, but before contracts for the disposal have been exchanged. The landlord may choose to serve a notice under this section on the qualifying tenants, and thereupon the Act shall cease to apply to the proposed disposal—if, however, the landlord chooses not to serve such a notice, then the Act will continue to apply to the proposed disposal as provided by section 10(3).

## 4-268
### Serve
See para 4-82 and para 4-273.

## 4-269
### Notice
See para 4-83.

## 4-270
### Qualifying tenants
See para 4-37.

## 4-271
### Constituent flats (section 5)
See para 4-160.

## 4-272

10(2)  A landlord who has not served such a notice on all of the qualifying tenants of the constituent flats shall nevertheless be treated as having duly served a notice under subsection (1)—

(a) if he has served such a notice on not less than 90% of those tenants, or
(b) where those qualifying tenants number less than ten, if he has served such a notice on all but one of them.

## 4-273

Section 10(2) provides that, even where the landlord omits to serve notice on one qualifying tenant (where there are fewer than 10 qualifying tenants) or omits to serve notices on 10% of the qualifying tenants (where there are 10 or more qualifying tenants), the landlord will nevertheless be treated as having served notice pursuant to section 10(1).

## 4-274

10(3)  Where the landlord is entitled to serve a notice under subsection (1) but does not do so, this Part shall continue to have effect in relation to the disposal in question as if the premises in question were still premises to which this Part applies.

## 4-275
See para 4-267. It seems very odd that premises that fall outside the 1987 Act are nonetheless deemed still to fall within the Act simply because the landlord has not bothered to serve a notice under section 10(1). As a result, the tenants can continue to exercise rights under an Act that no longer applies.

## 4-276
> 10(4) The above provisions of this section do not apply after a binding contract for the disposal of the protected interest—
>
> (a) has been entered into by the landlord and the nominated person, or
> (b) has otherwise come into existence between the landlord and the nominated person by virtue of any provision of this Part.

## 4-277
Once contracts for the proposed disposal have been exchanged between the landlord and the nominated person, the law of contract (including remedies relating to failure to complete, such as specific performance, damages, etc) applies to that contract and the 1987 Act (and its applicability or otherwise) becomes irrelevant (except where the landlord rescinds the contract under section 10(5)).

## 4-278
### Has otherwise come into existence
See para 4-248.

## 4-279
> 10(5) Where a binding contract for the disposal of the protected interest has been entered into between the landlord and the nominated person but it has been lawfully rescinded by the landlord, the landlord may, during the period of 12 months beginning with the date of the rescission of the contract, dispose of that interest to such person (and on such terms) as he thinks fit.

## 4-280
Where a buyer commits a serious breach of contract (for example, failure to complete the purchase following service of a notice to

complete that makes time of the essence in relation to the completion date), a seller will normally be entitled to rescind the contract whereupon the contract will come to an end without completion of the disposal having taken place.

Section 10(5) provides that where the landlord lawfully rescinds the contract that was entered into with the nominated person, the landlord will then, for a period of 12 months from the date of rescission, be free to dispose of the relevant interest to anybody and on any terms. This is the only situation where the landlord is free to dispose on any terms — in all other situations where the sale to the nominated person does not take place, the landlord can only dispose on the original terms and at no lower price than previously proposed (see section 7 above). It is odd that section 10(5) does not expressly state that the landlord will be free to dispose without needing to comply with the 1987 Act, but this would seem to be implied by the wording used.

### 4-281
#### 10A. Offence of failure to comply with requirements of Part I.

10A(1) A landlord commits an offence if, without reasonable excuse, he makes a relevant disposal affecting premises to which this Part applies—

(a) without having first complied with the requirements of section 5 as regards the service of notices on the qualifying tenants of flats contained in the premises, or
(b) in contravention of any prohibition or restriction imposed by sections 6 to 10.

### 4-282

Section 10A(1) makes it a criminal offence for the landlord to make a relevant disposal of premises within the 1987 Act, without complying with the requirements of section 5 (ie to serve offer notices on the qualifying tenants) and sections 6–10 (ie not to proceed with the disposal to a third party whilst the tenants/nominated person still have rights under the Act).

Note that where the landlord is a body corporate, a separate offence may be committed by each of those managing the body corporate under section 10A(3).

The disposal, however, is effective — see para 4-292; but the tenants have rights against the buyer under sections 11–17.

## 4-283
### Without reasonable excuse
There appears to be no case law on the meaning of this expression. See para 4-204.

## 4-284
### Relevant disposal
See para 4-47.

## 4-285
> 10A(2) A person guilty of an offence under this section is liable on summary conviction to a fine not exceeding level 5 on the standard scale.

## 4-286
A fine at level 5 on the standard scale is currently limited to a maximum of £5,000: see the Civil Justice Act 1982, section 37(2).

## 4-287
> 10A(3) Where an offence under this section committed by a body corporate is proved—
>
> (a) to have been committed with the consent or connivance of a director, manager, secretary or other similar officer of the body corporate, or a person purporting to act in such a capacity, or
> (b) to be due to any neglect on the part of such an officer or person, he, as well as the body corporate, is guilty of the offence and liable to be proceeded against and punished accordingly.
>
> Where the affairs of a body corporate are managed by its members, the above provision applies in relation to the acts and defaults of a member in connection with his functions of management as if he were a director of the body corporate.

## 4-288
Note that there is no reference to the landlord's external lawyers, surveyors or other professional advisors being guilty of an offence.

Where a criminal offence is committed by the landlord, however, the Proceeds of Crime Act 2002 may be relevant.

### 4-289
10A(4) Proceedings for an offence under this section may be brought by a local housing authority (within the meaning of section 1 of the Housing Act 1985).

### 4-290
Section 1 of the Housing Act 1985 defines "local housing authority" as:

> a district council, a London borough council, the Common Council of the City of London a Welsh county council or county borough council or the Council of the Isles of Scilly.

### 4-291
10A(5) Nothing in this section affects the validity of the disposal.

### 4-292
As previously stated, the disposal between the landlord and third party remains valid. The tenants, however, have rights against the third party under sections 11–17 of the 1987 Act.

### 4-293
**11. Circumstances in which tenants' rights enforceable against purchaser**

11(1) The following provisions of this Part apply where a landlord has made a relevant disposal affecting premises to which at the time of the disposal this Part applied ( 'the original disposal'), and either—

(a) no notice was served by the landlord under section 5 with respect to that disposal, or
(b) the disposal was made in contravention of any provision of sections 6 to 10,

and the premises are still premises to which this Part applies.

## 4-294

Section 11(1) introduces the rights enjoyed by the qualifying tenants against a purchaser who has taken a disposal that was made by the former landlord in breach of the former landlord's obligations under the 1987 Act. The landlord's obligation under section 5 is, briefly, to serve offer notices on the qualifying tenants, and the landlord's obligation under sections 6–10 is, briefly, not to proceed with the disposal to a third party whilst the tenants/nominated person still have rights under the Act.

Section 11(2) then specifies which section of the Act is relevant, depending on the nature of the unauthorised disposal that has been made.

In order for the qualifying tenants to have such rights, all of the following circumstances must exist:

- The landlord must have made a relevant disposal

- The premises affected by the disposal fell within the 1987 Act at the time of the disposal

- The landlord is in breach of obligations under either section 5 or sections 6–10

- The premises affected by the disposal still fall within the 1987 Act.

It should be noted that, although there are time limits within which the tenants must exercise their rights, these time limits do not start running until the purchaser has (i) notified the tenants of the disposal and (ii) notified the tenants of their rights in relation to that disposal. A purchaser is therefore caught on the horns of a dilemma: on the one hand, notifying the tenants of their rights may well lead them to exercise those rights against the purchaser (so that the purchaser may lose the building) but on the other hand, failure to notify the tenants of their rights means that the purchaser remains forever at risk of losing the building, as time does not start to run against the tenants: in the case of *Green* v *Westleigh Properties Ltd* (see para 4-307 for full case report), a tenant successfully claimed for specific performance of its demand that the freehold reversion be transferred to the tenants some 12 years after the original unauthorised disposal had taken place.

It should also be noted that, once an unauthorised disposal has taken place, the premises effectively become "tainted", in as much as the tenants' rights to seize the premises apply not just against the original third party purchaser but (by virtue of section 16) also against any subsequent purchaser, even though the disposal from original third party purchaser to subsequent purchaser may be made in compliance with the 1987 Act.

**4-295
Landlord**
See para 4-29.

**4-296
Relevant disposal**
See para 4-47.

**4-297
Affecting premises**
See para 4-04.

**4-298
Premises**
See para 4-05.

**4-299
To which at the time of the disposal**
See para 4-06.

**4-300**
11(2)  In those circumstances the requisite majority of the qualifying tenants of the flats contained in the premises affected by the relevant disposal (the 'constituent flats') have the rights conferred by the following provisions—

section 11A (right to information as to terms of disposal, &c.),
section 12A (right of qualifying tenants to take benefit of contract),
section 12B (right of qualifying tenants to compel sale, &c. by purchaser), and
section 12C (right of qualifying tenants to compel grant of new tenancy by superior landlord).

## 4-301
Section 11(2) sets out the various rights enjoyed by the tenants where a disposal in breach of the 1987 Act has taken place.

## 4-302
### Requisite majority of qualifying tenants
See para 4-99.

## 4-303
### Constituent flats (section 11)
See para 4-309.

## 4-304
> 11(3)  In those sections the transferee under the original disposal (or, in the case of the surrender of a tenancy, the superior landlord) is referred to as 'the purchaser'.
>
> This shall not be read as restricting the operation of those provisions to disposals for consideration.

## 4-305
### Original disposal
Section 11(1) defines this as a relevant disposal affecting premises to which, at the time of the disposal, the 1987 Act applied. However, in practice, the expression will only be relevant where the landlord has made a relevant disposal without complying with the Act.

## 4-306
> **11A.  Right to information as to terms of disposal, &c.**
>
> 11A(1)  The requisite majority of qualifying tenants of the constituent flats may serve a notice on the purchaser requiring him —
>
> (a) to give particulars of the terms on which the original disposal was made (including the deposit and consideration required) and the date on which it was made, and
> (b) where the disposal consisted of entering into a contract, to provide a copy of the contract.

11A(2) The notice must specify the name and address of the person to whom (on behalf of the tenants) the particulars are to be given, or the copy of the contract provided.

11A(3) Any notice under this section must be served before the end of the period of four months beginning with the date by which—

(a) notices under section 3A of the Landlord and Tenant Act 1985 (duty of new landlord to inform tenants of rights) relating to the original disposal, or

(b) where that section does not apply, documents of any other description—

 (i) indicating that the original disposal has taken place, and

 (ii) alerting the tenants to the existence of their rights under this Part and the time within which any such rights must be exercised,

have been served on the requisite majority of qualifying tenants of the constituent flats.

11A(4) A person served with a notice under this section shall comply with it within the period of one month beginning with the date on which it is served on him.

## 4-307

Where the landlord has made a relevant disposal of premises falling within the 1987 Act and has failed to comply with the Act in relation to the service of section 5 notices and/or in relation to any subsequent requirements, the requisite majority of qualifying tenants have rights against the purchaser to obtain information in relation to the disposal and then to step into the purchaser's shoes under the contract, or compel the purchaser to transfer the property to their nominated purchaser, or compel the grant of a new tenancy, whichever is applicable in the circumstances. Section 11A deals with tenants' rights to obtain information.

The requisite majority of qualifying tenants may serve notice on the purchaser, requiring particulars of the terms on which the disposal was made (including deposit and consideration required) and the date of the disposal, which the purchaser must comply with (section 11A(4)) within one month of service of that notice. There is no specified form for the notice, but it must:

- be served (section 11A(3)) before the end of the four-month period beginning with the date by which notices under section 3A of the Landlord and Tenant Act 1985 relating to the disposal have been served on the requisite majority of the qualifying tenants or, where section 3A does not apply, beginning on the date on which the requisite majority of qualifying tenants are served with documents indicating that the disposal has taken place and alerting them to the existence of their rights under the Act and the time within which such rights must be exercised

- be in writing (section 54(1))

- state the names of the persons on whose behalf it is served and the addresses of their flats (section 54(2))

- state the name and address of the person to whom the items requested by the notice are to be given, on behalf of the tenants (section 11A(2)).

The requisite majority of qualifying tenants then have a period of six months from the date on which the purchaser complied with the tenants' notice seeking information (or six months from the date when the tenants were notified of the disposal, if the tenants did not serve a notice seeking information) in which to serve a purchase notice, electing to take the benefit of the contract (section 12A), the property (section 12B), or the tenancy (section 12C), whichever is appropriate in the particular case.

The above provisions have given rise to disputes between landlord/purchaser and tenants, and there have been a number of cases dealing with these provisions.

The case of *Savva v Galway-Cooper* [2005] EWCA Civ 1068; [2005] 28 EG 120 (CS) concerned a building containing four flats, each of which was let to a qualifying tenant or tenants. The landlord had been granted two leases by the previous freeholder; one of the roof space and the other of the front garden, and section 5 notices had not been served. Accordingly, the requisite majority of qualifying tenants wished to exercise their rights to have the two leases assigned to them.

The case turned on the question of what the landlord/purchaser needed to do in order to serve, on the requisite majority of qualifying tenants, documents indicating that the disposal had taken place and alerting them to the existence of their rights under the Act and the time within which such rights must be exercised, so that time would start running against the tenants in relation to service of a purchase notice.

In the case of the first flat, service had been effected by a letter from the tenant's own solicitor to the tenant himself. In the case of the second flat, service had never been effected. In the case of the third flat, the tenant's solicitor had obtained office copy entries from the Land Registry in which the two leases were mentioned in a schedule. In the case of the fourth flat, successive tenants had been sent copies of one of the leases, and the current tenant had been given particulars of the second lease by the defendant.

Perhaps not surprisingly, the Court of Appeal held that service by the landlord was a formal step, which could not be done incidentally. Service means "the supply of some document in relationship (actual or prospective) between parties with distinct legal interests ... receipt of office copy entries from the Land Registry, or a letter from one's own solicitor or, even less, information from reading a newspaper" is not service. There had not been the requisite service in relation to any of the four flats.

See also the case of *Staszewski* v *Maribella Ltd* [1998] 1 EGLR 34, where the Court of Appeal held that time had not started to run against the tenants, because information provided by the landlord was inaccurate and the landlord had not made it clear why he was providing the information. All terms relating to money (whether or not in the contract) must be notified to the tenants; this was confirmed in the case of *Michaels* v *Harley House (Marylebone) Ltd* [1998] EWCA Civ 1714; [1998] EGCS 159, where the third party purchaser's failure to inform the tenants of the details of loans made to enable the purchase of the building meant that it had not provided a valid response to the tenants' request for information.

In the case of *M25 Group Ltd* v *Tudor*, it was the tenants who had failed to comply with the strict requirements of the Act. The requisite majority of qualifying tenants served a notice on the purchaser of the freehold reversion (the sale to him having been in breach of the Act), seeking information in relation to the disposal. The notice named the tenants but did not state the addresses of the flats of which they were the qualifying tenants, as required by section 54(2) of the Act.

The Court of Appeal held that the notice had not been invalidated by the absence of the flat addresses. The Court distinguished between substantive provisions of a statute (which in this case were those conferring the right to acquire the freehold) and secondary or "machinery" provisions (which in this case were the notice requirements and the formal requirements of section 54 of the Act, including the requirement for the addresses of the flats of the qualifying tenants). The machinery provisions could then be divided into essential machinery (such as the requirement for a notice, and the requirement to indicate who the notice is from) and non-essential, merely supportive machinery (such as the requirement to state addresses). The breach of the merely supportive machinery did not invalidate the notice.

In the recent case of *Green v Westleigh Properties Ltd*, following an unauthorised disposal by the freeholder, the County Court ordered the tenant to serve a purchase notice under section 12 of the 1987 Act. The tenant purported to do so, but the notice contained a number of errors —it referred to the Housing Act 1987 (which did not exist), it quoted section 12A of the 1987 Act (which was inapplicable), it did not specify the addresses of the flats of which the tenants were qualifying tenants, and it bore the wrong date.

Despite these errors, the High Court held that a valid purchase notice had been served. The imperative requirements for this notice were that it should be in writing, served on the new landlord in time, and that it should give adequate notice of the requirement of the qualifying tenants to have the interest in the building. Based on the principles of *Mannai Investment Co Ltd v Eagle Star Assurance Co Ltd*, the reasonable recipient of the notice could have been left in no doubt that the appellant and the other tenant were giving notice of the requirement to have the freehold of the building transferred to them, despite incorrect statutory references.

## 4-308
### Requisite majority of qualifying tenants (section 11A)

Section 18A of the 1987 Act defines "requisite majority of qualifying tenants", for the purposes of section 11A, to mean qualifying tenants having more than 50% of the available votes for flats in the premises, as at the date of service of notice under section 11A. Section 11A provides for the service of two notices: one by the landlord, giving details of the disposal and the tenants' rights in relation to it, and one by the tenants exercising their right to receive information. It is not clear which of these notices is applicable for the purposes of this definition.

Each flat occupied by a qualifying tenant is deemed to have a single vote for this purpose: section 18A(3). The Act does not deal with what would happen if a tenant of a particular flat in fact comprised more than one individual, and each individual voted differently.

Note that the requisite majority of qualifying tenants must have more than 50% of the available votes. So, more than half of the number of qualifying tenants is required for this majority. This means that in the case of a building containing only two qualifying tenants, both are required to constitute a requisite majority. In the case of a building containing an even number of flats let to qualifying tenants, one more than half of those qualifying tenants is required to constitute a requisite majority, as the following table shows:

| Number of flats let to qualifying tenants in building | Number of qualifying tenants required to constitute "requisite majority" |
| --- | --- |
| 2 | 2 |
| 3 | 2 |
| 4 | 3 |
| 5 | 3 |
| 6 | 4 |
| 7 | 4 |
| 8 | 5 |
| 9 | 5 |
| 10 | 6 |

See also para 4-316.

## 4-309
### Constituent flats (section 11)

This expression is defined in section 11(2) to mean "the flats contained in the premises affected by the relevant disposal". "The premises" for this purpose are the premises affected by the relevant disposal—see para 4-04.

Note that section 5(1) also defines "constituent flats" but slightly differently from the definition in section 11(2). The section 5(1) definition simply refers to "the flats contained in the premises", whereas the section 11(2) definition refers to "the flats contained in the premises affected by the relevant disposal". The question of whether or not a disposal "affects" premises was considered in the case of *Dartmouth Court Blackheath Ltd* v *Berisworth Ltd*—see para 4-04.

## 4-310
## Serve
See para 4-82.

## 4-311
## Notice
See para 4-83.

## 4-312
## Purchaser
Section 11(3) provides that any reference to "the purchaser" in sections 11A, 12A, 12B and 12C shall mean the transferee under the original disposal (as defined in section 11(3) above) and, in the case of a disposal by way of the surrender of a lease, shall mean the superior landlord. Despite the use of the word "purchaser", section 11(3) clarifies the fact that absence of consideration will not prevent a disposal from being a relevant disposal.

## 4-313
## Original disposal
See para 4-305.

## 4-314
## Before the end of the period of four months
Section 11A(3) specifies a four-month period from the date of service of the notices under section 3A of the Landlord and Tenant Act 1985, where applicable, after which the tenants' right to obtain information about the disposal is lost. However, section 3A states that, in the notice served on the tenants under section 3A, the landlord must inform the tenants that the timing for exercise of any of their rights commences on the date of **receipt** of the notice, not on the date on which it is served. Thus, section 3A obliges the landlord to give information to the tenants about the timing for the exercise of their rights that is incorrect (except where service and receipt of the notice occurs on the same date).

Clearly, a landlord will wish to comply with the obligations under section 3A of the 1985 Act, since otherwise the landlord will commit a criminal offence. If, however, the tenants act in reliance on the landlord's section 3A notice and exercise their rights within the period of four months after receipt but after the expiry of the period of four months after service of the notice, they will technically be out of time for exercising their rights. It seems likely, however, that a court would hold that the landlord is estopped from taking the point, given that the tenants have acted in reliance on the information supplied by the landlord.

Where, however, the relevant disposal did not result in a change of landlord (such as an exchange of contracts), neither section 3 nor section 3A of the 1985 Act will apply. In such a case, the period for exercise of the tenant's rights to receive information about the disposal begins with the date on which documents have been served on the requisite majority of qualifying tenants indicating that the disposal has taken place and alerting them to their rights under the 1987 Act (and the timing for exercise of such rights). See also para 4-316.

### 4-315
### Section 3A of the Landlord and Tenant Act 1985

Section 3 of the Landlord and Tenant Act 1985 provides that a purchaser of the landlord's interest in a building must notify all the residential tenants in that building that he or she is now their landlord, and it is a criminal offence to fail to do so. Section 3A of that Act then provides that, where the landlord is obliged to give a section 3 notice and a relevant disposal has taken place, the landlord must also serve notice on all the qualifying tenants in the building, notifying them of the relevant disposal, that it was a disposal caught by the 1987 Act and of the rights that they may have under the 1987 Act. Rather oddly, the obligation under section 3A applies whether or not the relevant disposal was made in breach of the 1987 Act requirements, so that the qualifying tenants must be informed of the rights that they may have, despite the fact that they may in practice have no such rights in relation to that disposal.

### 4-316
### Served on the requisite majority of qualifying tenants

We have already noted that, where the disposal has resulted in a change of landlord, all of the qualifying tenants must be notified of the disposal and their rights in relation to it. Where, however, the disposal has not resulted in a change of landlord (as in the case of an exchange of contracts), it seems most surprising that only the requisite majority of qualifying tenants need to be notified of the disposal and their rights in relation to it under section 11A(3).

The requisite majority of qualifying tenants is defined in section 18A to be qualifying tenants having more than 50% of the available votes. Where, therefore, the building that was the subject of the disposal contains 20 flats, each let to a qualifying tenant, only 11 of those tenants are entitled to receive information from the landlord under section 11A(3).

There is no guidance on how the landlord is expected to decide which 11 tenants out of the total number of 20 should be served with the information: clearly the landlord is free to serve all 20 tenants if the landlord so wishes, but is under no obligation to serve more than 11. If the landlord serves only 11 tenants, all of them must serve notice on the landlord, exercising their rights in relation to the disposal, otherwise the rights will not have been exercised by a sufficient number of qualifying tenants to constitute a requisite majority. Further complications may arise where some tenants are served on one date and others on a different date, as the four-month period runs from the date on which the first 11 tenants are served.

## 4-317
**12A. Right of qualifying tenants to take benefit of contract.**

12A(1) Where the original disposal consisted of entering into a contract, the requisite majority of qualifying tenants of the constituent flats may by notice to the landlord elect that the contract shall have effect as if entered into not with the purchaser but with a person or persons nominated for the purposes of this section by the requisite majority of qualifying tenants of the constituent flats.

## 4-318
Section 12A(1) gives the tenants the right, by serving notice on the landlord to that effect, to elect that the nominated person shall step into the shoes of the third party purchaser under the contract entered into, in breach of the 1987 Act, between the nominated person and the landlord. Note that there is no requirement for the nominated person to be the same person who was to receive the information sought under section 11A.

## 4-319
## Original disposal
See para 4-305.

## 4-320
## Entering into a contract
Section 12A(1), by commencing with the words: "Where the original disposal consisted of entering into a contract", implies that this is the section that is relevant where the original disposal consisted of entering into a contract. It should be noted, however, that where completion under that contract has taken place before the tenants have exercised their rights, section 12B(1)(a) appears to be the relevant section: see para 4-342.

It is crucial that the tenants serve notice under the correct section: see *Kensington Heights Commercial Co Ltd* v *Campden Hill Developments Ltd* [2007] EWCA Civ 245; [2007] 1 EGLR 130. A full report of this case can be found in para 4-396.

It is clear that a disposal by way of contract to be completed by way of conveyance (section 5A) and a disposal at public auction (section 5B) constitute disposals which consist of entering into a contract. Similarly, it is clear that a disposal by way of conveyance not preceded by an exchange of contracts is not a disposal which consists of entering into a contract (and accordingly will fall within section 12B(1)(b) rather than section 12A(1)).

The position is less clear, however, in the case of an option or right of pre-emption. Normally, the grantor will grant an option or right of pre-emption directly to the grantee, without this being preceded by a contract for the grant. An option or right of pre-emption is not itself a contract: see *Spiro* v *Glencrown Properties Ltd* [1991] 1 EGLR 185, in which the Court described an option as a relationship sui generis which is neither an offer nor a contract. Upon the exercise of the option or right of pre-emption, however, a contract for the disposal of an interest in land will automatically be created.

It appears therefore that, where an option or right of pre-emption has been granted, not preceded by a contract, and the landlord has not complied with the obligations under section 5 of the 1987 Act in relation to that grant, the tenants should exercise their rights under section 12B(1)(b) rather than under section 12A(1).

## 4-321
## Requisite majority of qualifying tenants (section 12A/B/C)

Section 18A of the 1987 Act defines "requisite majority of qualifying tenants", for the purposes of sections 12A, 12B and 12C, to mean qualifying tenants having more than 50% of the available votes for flats in the premises, as at the date of service of the tenants' notice under each of those sections.

Note that the requisite majority of qualifying tenants must have more than 50% of the available votes. For more details, see para 4-99.

## 4-322
## Constituent flats (section 11)
See para 4-309.

## 4-323
## Notice
See para 4-83. See also para 4-82 in relation to service of the notice.

## 4-324
## Landlord
See para 4-29.

## 4-325
## The contract shall have effect... (section 12A(1))

The wording of section 12A(1) implies that the legal effect of the contract between the landlord and the third party purchaser is altered so that for all purposes, the contract now takes effect between landlord and nominated person rather than landlord and purchaser. This begs the question: if the third party purchaser no longer has the benefit of the contract, does the third party purchaser have any remedy against the landlord? This will depend on any representations made by the landlord to the third party, and the third party may well be able to bring a claim for misrepresentation against the landlord if, for example, the landlord confirmed in replies to enquiries before contract that the property did not fall within the 1987 Act. As the purchaser no longer has the benefit of the contract, it would appear that the purchaser cannot bring a claim under that contract, for example for breach of warranty.

## 4-326
### Purchaser
See para 4-312 and para 4-458.

## 4-327
### Person or persons nominated
See para 4-130.

## 4-328

> 12A(2) Any such notice must be served before the end of the period of six months beginning—
>
> > (a) if a notice was served on the purchaser under section 11A (right to information as to terms of disposal, &c.), with the date on which the purchaser complied with that notice;
> > (b) in any other case, with the date by which documents of any description —
> > > (i) indicating that the original disposal took place, and
> > > (ii) alerting the tenants to the existence of their rights under this Part and the time within which any such rights must be exercised,
>
> have been served on the requisite majority of qualifying tenants of the constituent flats.

## 4-329

Section 12A(2) prescribes a six-month period within which the tenants must serve notice on the landlord, if they are to exercise their rights in relation to the contract that was entered into in breach of the 1987 Act. The date on which this six-month period commences depends on whether or not the tenants first exercised their rights under section 11A to obtain from the third party purchaser information about the disposal. If they did, then the period starts on the day on which the purchaser provided the information, but if they did not then the period starts on the day on which documents were served on the tenants, indicating that the disposal had taken place and alerting the tenants to their rights under the Act.

## 4-330
### Served
The case of *Savva* v *Galway-Cooper* deals with the meaning of service in this context, see para 4-307 for the full case report.

## 4-331
### Served on the requisite majority of qualifying tenants
See para 4-316.

## 4-332
> 12A(3) The notice shall not have effect as mentioned in subsection (1) unless the nominated person—
>
> (a) fulfils any requirements as to the deposit required on entering into the contract, and
> (b) fulfils any other conditions required to be fulfilled by the purchaser on entering into the contract.

## 4-333
Although it is the requisite majority of qualifying tenants that must serve notice under section 12A, it is the nominated person who must fulfil the requirements in section 12A(3), without which the notice served by the requisite majority of qualifying tenants will be of no effect.

## 4-334
### Fulfils any requirements as to the deposit
Section 12A(1), as previously mentioned, provides that the contract originally entered into between landlord and third party purchaser shall have effect as though it had been entered into between the landlord and the nominated person. Assuming that any deposit payable on entering into the contract has already been paid by the third party purchaser to the landlord, section 12A(3)(a) raises a number of issues:

- Is the nominated person required to pay a second deposit to the landlord? The natural meaning of "fulfils any requirements as to the deposit" would seem to suggest that the nominated person is expected to pay a (second) deposit to the landlord, but if the nominated person is treated as having entered into the contract in place of the third party purchaser, and a deposit has been paid, why does the nominated person have to pay it again?

- If the nominated person pays a (second) deposit to the landlord, can the third party purchaser compel the landlord to repay the original deposit?

Natural justice would suggest that the nominated person should pay a deposit and the landlord should repay the deposit to the third party purchaser, but the Act does not provide for this.

### 4-335
### On entering into the contract
Does this mean that the nominated person must pay the deposit and fulfil any other conditions when the nominated person steps into the shoes of the third party purchaser as purchaser under the contract? Or do the words "on entering into the contract" qualify the date on which the deposit had to be paid under the contract, and the date on which the conditions had to be fulfilled under the contract (rather than referring to the date on which the nominated person must pay/comply)?

If the former is correct, the nominated person must fulfil any conditions of the contract not later than the date on which the tenants serve their section 12A(1) notice—which is unworkable if the contract is, for example, conditional on the superior landlord's consent. If the latter is correct, the nominated person need only pay/comply to the extent that payment/compliance is required on exchange of contracts.

It is submitted that the latter is more likely to be correct, because the words "on entering into the contract" are repeated in sub-subsections (a) and (b) of section 12A(3), rather than simply appearing at the end of subsection (3). If so, it should be possible for the nominated person to pay/comply on a different (later) date from that on which the tenants serve their section 12A(1) notice. Presumably in such a case the tenants' notice would be retrospectively validated for the purposes of subsection (3) when the payment/compliance takes place.

### 4-336
12A(4) Unless otherwise agreed, any time limit in the contract as it has effect by virtue of a notice under this section shall start to run again on the service of that notice; and nothing in the contract as it has effect by virtue of a notice under this section shall require

the nominated person to complete the purchase before the end of the period of 28 days beginning with the day on which he is deemed to have entered into the contract.

## 4-337

Having stepped into the shoes of the third party purchaser under section 12A(1) above, the nominated person is obliged to comply with the obligations under the contract. However, a substantial period of time may have elapsed before the nominated person is substituted for the third party purchaser. For this reason, the nominated person cannot be expected to comply with the same time limits, or to complete the purchase on the same completion date, as the third party purchaser.

This section therefore provides (i) that any time limit in the contract will start to run again on the service of the tenants' notice under section 12A(1); and (ii) that the completion date cannot occur less than 28 days after "the day on which [the nominated person] is deemed to have entered into the contract" (see para 4-338).

## 4-338
## The day on which [the nominated person] is deemed to have entered into the contract

Section 12A(1) states that the contract shall have effect as if the nominated person had been the contracting party (rather than the third party purchaser), which suggests that the nominated person is deemed to have entered into the contract on the date on which the tenants served notice under section 12A(1).

Where notice is served on one day and the requirements of section 12A(3) are satisfied on a different day (see para 4-335), the 1987 Act does not make clear on which of the two dates the contract is deemed to have been entered into. Section 12A(4) provides for time limits in the contract to start to run again on the date on which the tenants serve notice under section 12A(1). There is no reference to compliance with any conditions, nor to the fact that this may take place after service of notice.

Furthermore, section 12A(1) states that the contract shall have effect as if the nominated person had entered into it. The contract was created when it was entered into between the landlord and the third party purchaser. Could this mean that the nominated person is deemed to have entered into the contract on that date? If so, the provision in section 12A(4) that the nominated person cannot be required to complete less than 28 days after being deemed to have entered into the contract may well give rise to difficulties.

## 4-339

(5) Where the original disposal related to other property in addition to the premises to which this Part applied at the time of the disposal—

(a) a notice under this section has effect only in relation to the premises to which this Part applied at the time of the original disposal, and
(b) the terms of the contract shall have effect with any necessary modifications.

In such a case the notice under this section may specify the subject-matter of the disposal, and the terms on which the disposal is to be made (whether doing so expressly or by reference to the original disposal), or may provide for that estate or interest, or any such terms, to be determined by a leasehold valuation tribunal.

## 4-340

Section 12A(5) deals with the situation where the original disposal by the landlord to the third party purchaser related both to premises within the 1987 Act and to premises outside the Act. In such a case, the tenants can only exercise their rights in relation to those premises that fall within the Act, and the original contract, to which the nominated person is treated as becoming a party in place of the third party purchaser, is amended as necessary. Either the tenants' notice can specify how the contract will work between the landlord and the nominated person, or the notice can provide for this to be determined by a leasehold valuation tribunal.

It is interesting to contrast this provision with section 5(3), which requires the landlord to sever a transaction that relates to more than one building. There appears to be nothing in the Act to require a landlord to sever a transaction that relates to both premises within the Act and premises outside the Act (it may be that the meaning of "building" precludes the inclusion of premises within the Act and premises outside the Act within a single building—see para 4-17). There also appears to be nothing in sections 11–17 that requires a disposal relating to more than one building to be severed.

## 4-341
**12B. Right of qualifying tenants to compel sale, &c. by purchaser.**

12B(1) This section applies where —

(a) the original disposal consisted of entering into a contract and no notice has been served under section 12A (right of qualifying tenants to take benefit of contract), or

(b) the original disposal did not consist of entering into a contract.

12B(2) The requisite majority of qualifying tenants of the constituent flats may serve a notice (a 'purchase notice') on the purchaser requiring him to dispose of the estate or interest that was the subject-matter of the original disposal, on the terms on which it was made (including those relating to the consideration payable), to a person or persons nominated for the purposes of this section by any such majority of qualifying tenants of those flats.

## 4-342
Section 12B(2) gives the tenants the right, by serving notice on the third party purchaser to that effect, to require the third party purchaser to hand over to the nominated person the subject matter of the purchase from the landlord on the terms on which the third party purchased it. Note that there is no requirement for the nominated person to be the same person who was to receive the information sought under section 11A.

In the same way that section 12A appears to be intended to apply where contracts have been exchanged between landlord and third party purchaser but completion has not taken place, section 12B appears to be intended to apply only where completion has taken place in relation to the transaction that was the subject of the original contract between landlord and third party purchaser, or where there was never a preliminary contract at all. The fact that the tenants' notice is defined in section 12B(2) as a "purchase notice" supports this interpretation of section 12B, as does the fact that section 12B(3)(b)(i) contemplates that a change of landlord may have taken place following the disposal, whereas section 12A(2)(b) does not. The comments of Lawrence Collins LJ in *Kensington Heights Commercial Co Ltd* v *Campden Hill Developments Ltd* at para 84 are also consistent with this approach:

> Section 12A ... applies to cases 'where the original disposal consisted of entering into a contract' and deals with the case where the contract has

not been completed. It contemplates that a notice is served by the tenants on the landlord electing that the contract shall have effect as if entered into not by the purchaser but by the person or persons nominated by the tenants. This is only workable if the contract has not been completed.

A full report of this case can be found in para 4-396.

However, the wording of section 12B(1)(a) implies that, where there has been an exchange of contracts between landlord and purchaser, the tenants may choose whether to serve notice on their original landlord under section 12A(1) or on the third party purchaser under section 12B(2), and in either such case the nominated person will step into the shoes of the third party purchaser as purchaser under the contract.

There would, of course, be no point in the nominated person seeking to become the purchaser under a contract that has already been completed; either the contract will have merged on completion (in which case it no longer exists) or the subject matter of the contract will have already passed to the third party purchaser, so that only minimal benefit will remain in the contract itself. Where, however, completion has not yet taken place, and there is no prohibition on assignment of the benefit of the contract by the purchaser, it would be theoretically possible for the third party purchaser to assign to the nominated person the benefit of the contract that was entered into with the landlord, thus placing the nominated person in the same position as if a section 12A notice had been served on the landlord instead.

On balance, it seems preferable for the tenants to serve notice under section 12A(1) rather than under section 12B(2) where contracts have been exchanged with the third party purchaser but completion has not yet taken place. Note that under no circumstances can notice be served under both section 12A and section 12B, as section 12B(1)(a) precludes service of notice under that section where a section 12A notice has been served.

## 4-343
## Original disposal
See para 4-305.

## 4-344
## Entering into a contract
See para 4-320. See also para 4-342.

## 4-345
### Requisite majority of qualifying tenants (section 12A/B/C)
Section 18A of the 1987 Act defines "requisite majority of qualifying tenants", for the purposes of sections 12A, 12B and 12C, to mean qualifying tenants having more than 50% of the available votes for flats in the premises, as at the date of service of the tenants' notice under each of those sections.

Note that the requisite majority of qualifying tenants must have more than 50% of the available votes. For more details, see para 4-99.

## 4-346
### Constituent flats (section 11)
See para 4-309.

## 4-347
### Serve
See para 4-82.

## 4-348
### Notice
See para 4-83.

## 4-349
### Purchaser
See para 4-312 and para 4-458.

It is crucial to note that the tenants serve their notice on the purchaser in the case of section 12B, not the landlord as in the case of section 12A. Service on the wrong party can be fatal: see *Kensington Heights Commercial Co Ltd v Campden Hill Developments Ltd*. A full report of this case can be found in para 4-396.

## 4-350
### Person or persons nominated
See para 4-130.

## 4-351
12B(3) Any such notice must be served before the end of the period of six months beginning—

(a) if a notice was served on the purchaser under section 11A (right to information as to terms of disposal, &c.), with the date on which the purchaser complied with that notice;
(b) in any other case, with the date by which—
  (i) notices under section 3A of the Landlord and Tenant Act 1985 (duty of new landlord to inform tenants of rights) relating to the original disposal, or
  (ii) where that section does not apply, documents of any other description indicating that the original disposal has taken place, and alerting the tenants to the existence of their rights under this Part and the time within which any such rights must be exercised, have been served on the requisite majority of qualifying tenants of the constituent flats.

### 4-352
Section 12B(3) prescribes a six-month period within which the tenants must serve notice on the purchaser, if they are to exercise their rights in relation to the disposal that was made in breach of the 1987 Act. The date on which this six-month period commences depends on whether or not the tenants first exercised their rights under section 11A to obtain from the third party purchaser information about the disposal. If they did, then the period starts on the day on which the purchaser provided the information, but if they did not, then the period starts on the day on which notices under section 3A of the Landlord and Tenant Act 1985 relating to the disposal were served on the tenants or, where section 3A does not apply, the day on which documents were served on the tenants, indicating that the disposal had taken place and alerting the tenants to their rights under the Act.

### 4-353
### Served
The case of *Savva* v *Galway-Cooper* deals with the meaning of service in this context; see the case report in para 4-307.

### 4-354
### Served on the requisite majority of qualifying tenants
See para 4-316.

## 4-355
## Section 3A of the Landlord and Tenant Act 1985
See para 4-315.

## 4-356
12B(4) A purchase notice shall where the original disposal related to other property in addition to premises to which this Part applied at the time of the disposal—

    (a) require the purchaser only to make a disposal relating to those premises, and

    (b) require him to do so on the terms referred to in subsection (2) with any necessary modifications.

In such a case the purchase notice may specify the subject-matter of the disposal, and the terms on which the disposal is to be made (whether doing so expressly or by reference to the original disposal), or may provide for those matters to be determined by a leasehold valuation tribunal.

## 4-357
Section 12B(4) deals with the situation where the original disposal by the landlord to the third party purchaser related both to premises within the 1987 Act and to premises outside the Act. In such a case, the tenants can only exercise their rights in relation to those premises that fall within the Act, and the terms of the original disposal, on the basis of which the nominated person will acquire the subject matter of that disposal from the third party purchaser, are amended as necessary. Either the tenants' notice can specify the subject matter of the disposal and the terms on which it is to be made, or the notice can provide for this to be determined by a leasehold valuation tribunal.

## 4-358
### Purchase notice
Section 12B(2) defines the notice served by the tenants on the purchaser under that section as a "purchase notice".

## 4-359
12B(5) Where the property which the purchaser is required to dispose of in pursuance of the purchase notice has since the original disposal

*Mixed Use and Residential Tenants' Rights*

become subject to any charge or other incumbrance, then, unless the court by order directs otherwise—

(a) in the case of a charge to secure the payment of money or the performance of any other obligation by the purchaser or any other person, the instrument by virtue of which the property is disposed of by the purchaser to the person or persons nominated for the purposes of this section shall (subject to the provisions of Part I of Schedule 1) operate to discharge the property from that charge; and

(b) in the case of any other incumbrance, the property shall be so disposed of subject to the incumbrance but with a reduction in the consideration payable to the purchaser corresponding to the amount by which the existence of the incumbrance reduces the value of the property.

## 4-360

Section 12B(5) deals with the situation where the property that is to be acquired by the nominated person from the third party purchaser has been charged or otherwise encumbered after the purchaser bought it from the landlord. Unless the court orders otherwise, a charge to secure payment of money or performance of obligations will be discharged on purchase by the nominated person; however, any other incumbrance will remain, but the price paid by the nominated person will be reduced accordingly.

## 4-361
### Charge

There is no specific definition of "charge" in the 1987 Act. However, section 12B(6) states that mortgages and liens, but not rent charges, will fall within the provisions relating to charges (subject to any necessary modifications). Part I of Schedule 1 to the Act, which deals with redemption of charges, refers to both debentures and floating charges, although a floating charge in favour of trustees for debenture holders will not be discharged under the provisions of section 12B(5)(a).

## 4-362
### Incumbrance

There is no specific definition of "incumbrance" in the 1987 Act, but the reference to "any other incumbrance" in (b) above implies that the Act regards a mortgage, charge or lien as an incumbrance. Depending on the context, a lease or a third party right (such as a covenant or easement) could be an incumbrance.

Case law indicates that the following are incumbrances for the purposes of the 1987 Act:

- A lease of a single flat (but not a reversionary lease): see *Belvedere Court Management Ltd* v *Frogmore Developments Ltd* [1996] 1 EGLR 59. However, a reversionary lease was held to be an incumbrance (and the nominated purchaser was to take free of it, which seems odd) at first instance in the case of *Boyle* v *Hallstate Ltd* [2002] EWHC 972.

- A repairs notice that has been served under the Housing Act 1985: see *Gregory* v *Saddiq* [1991] 1 EGLR 237.

- A lease of roof space (not demised to the tenants): see *Englefield Court Tenants* v *Skeels* [1990] 2 EGLR 230.

Any incumbrance, the grant of which constitutes a relevant disposal for the purposes of the Act can, of course, be acquired by the tenants under section 12 of the Act. Accordingly, only incumbrances, the grant of which does not constitute a relevant disposal (such as the grant of an easement—see section 4(2)(a)(iii)), will cause a problem to the tenants, in which case they can reduce the price commensurately under section 12B(5)(b).

## 4-363
### Subject to the provisions of Part I of Schedule 1

Part I of Schedule 1 to the 1987 Act provides in detail for the redemption of a charge within section 12B(5)(a), ie a charge to secure the payment of money or the performance of any other obligation.

Although the wording of section 12B(5)(a) implies that the charge will be discharged, in fact the nominated person must (and indeed is under a duty to) pay the purchase price (or the appropriate part thereof required to repay the charge) to the chargee, otherwise the charge will remain in place: para 2(1) of Part I of Schedule 1. Provided that the nominated person does this, the transfer document made between the third party purchaser and the nominated person will operate as a discharge of the charge, even where the price is less than the amount outstanding under the charge: para 2(2).

In case of difficulty in obtaining a redemption figure, finding the chargee or persuading the chargee to accept an inadequate figure in discharge of the charge, the nominated person may pay the purchase price (or the part needed to redeem the charge, if known) into court: para 4. Where there is more than one charge over the property, payment must be made to the chargees according to their respective priorities: para 2(1).

A fixed charge in favour of trustees for debenture holders will be discharged under the provisions of the Schedule, but a floating charge will not be: paras 2(3) and 2(4). Presumably this is because the nominated person will acquire the property free from any floating charge, provided that it has not crystallised at the date of completion of the acquisition.

### 4-364

12B(6) Subsection (5)(a) and Part I of Schedule 1 apply, with any necessary modifications, to mortgages and liens as they apply to charges; but nothing in those provisions applies to a rentcharge.

### 4-365

See para 4-360.

### 4-366

12B(7) Where the property which the purchaser is required to dispose of in pursuance of the purchase notice has since the original disposal increased in monetary value owing to any change in circumstances (other than a change in the value of money), the amount of the consideration payable to the purchaser for the disposal by him of the property in pursuance of the purchase notice shall be the amount that might reasonably have been obtained on a corresponding disposal made on the open market at the time of the original disposal if the change in circumstances had already taken place.

### 4-367

It is apparent that there may be a considerable delay between the third party purchaser acquiring the property from the landlord and the nominated person completing the acquisition of the property from the third party purchaser—see para 4-294. Also, see *Green* v *Westleigh Properties Ltd*, in which the tenants acquired the property more than 12 years after the landlord had transferred it to the third party purchaser in breach of the 1987 Act. Where the value of the property has increased during that period, it would clearly be unfair if the nominated person were able to acquire the property at less than its revised value.

Section 12B(7), however, provides that the nominated person must pay the third party purchaser the increased value of the property "owing to a change in circumstances other than a change in the value of money". The property is revalued as at the date of its acquisition by the purchaser as if the change in circumstances had already taken place at that earlier date. The purchaser will not, therefore, get a price that reflects any increase in property values generally. Presumably, the reasoning behind this provision is that the purchaser has bought the property regardless of the landlord's failure to comply with the 1987 Act and so deserves to be penalised, but where the purchaser has actually improved the property, the purchaser should be compensated.

Section 12B(7) deals only with increases in the value of the property; where the property has declined in value since the purchaser bought it, either because of a decline in property prices or because of some act of the purchaser, there is nothing in the Act that allows the nominated person to pay a lower price than that originally paid by the purchaser to the landlord.

## 4-368

**12C. Rights of qualifying tenants to compel grant of new tenancy by superior landlord**

12C(1) This section applies where the original disposal consisted of the surrender by the landlord of a tenancy held by him ('the relevant tenancy').

12C(2) The requisite majority of qualifying tenants of the constituent flats may serve a notice on the purchaser requiring him to grant a new tenancy of the premises which were subject to the relevant tenancy, on the same terms as those of the relevant tenancy and so as to expire on the same date as that tenancy would have expired, to a person or persons nominated for the purposes of this section by any such majority of qualifying tenants of those flats.

## 4-369

Section 12C(2) gives the tenants the right, by serving notice on the third party purchaser to that effect, to require the third party purchaser to grant to the nominated person a new tenancy to replace the tenancy originally surrendered by the landlord in breach of the 1987 Act. The new tenancy will be on the same terms as the surrendered tenancy, and will expire on the same date. Note that there is no requirement for the nominated person to be the same person who was to receive the information sought under section 11A.

**4-370**
**Original disposal**
See para 4-305.

**4-371**
**Landlord**
See para 4-29.

**4-372**
**Requisite majority of qualifying tenants (section 12A/B/C)**
See para 4-321.

**4-373**
**Constituent flats (section 11)**
See para 4-309.

**4-374**
**Serve**
See para 4-82.

**4-375**
**Notice**
See para 4-83.

**4-376**
**Purchaser**
See para 4-312 and para 4-458.
 Note that the purchaser under section 12C will always be the superior landlord because the purchaser is the person to whom the landlord has surrendered the lease.

As in the case of section 12B, it is crucial that the tenants serve their notice on the purchaser and not on the landlord—see *Kensington Heights Commercial Co Ltd* v *Campden Hill Developments Ltd*. A full report of this case can be found in para 4-396.

## 4-377
### Relevant tenancy
This is defined in section 12C(1) as the tenancy surrendered by the landlord to the third party purchaser under the original disposal.

## 4-378
### Person or persons nominated
See para 4-130.

## 4-379
12C(3) Any such notice must be served before the end of the period of six months beginning—

    (a) if a notice was served on the purchaser under section 11A (right to information as to terms of disposal, &c.), with the date on which the purchaser complied with that notice;

    (b) in any other case, with the date by which documents of any description—

        (i) indicating that the original disposal has taken place, and

        (ii) alerting the tenants to the existence of their rights under this Part and the time within which any such rights must be exercised,

have been served on the requisite majority of qualifying tenants of the constituent flats.

## 4-380
Section 12C(3) prescribes a six-month period within which the tenants must serve notice on the third party purchaser, if they are to exercise their rights in relation to the surrender that was made in breach of the 1987 Act. The date on which this six-month period commences depends on whether or not the tenants first exercised their rights under section 11A to obtain information about the disposal from the third party purchaser. If they did, then the period starts on the day on

which the purchaser provided the information, but if they did not, then the period starts on the day on which the documents were served on the tenants, indicating that the disposal had taken place and alerting the tenants to their rights under the Act.

### 4-381
### Served
The case of *Savva* v *Galway-Cooper* deals with the meaning of service in this context; see the case report in para 4-307.

### 4-382
### Served on the requisite majority of qualifying tenants
See para 4-308.

### 4-383
12C(4) If the purchaser paid any amount to the landlord as consideration for the surrender by him of that tenancy, the nominated person shall pay that amount to the purchaser.

### 4-384
It is interesting to contrast section 12C(4), which clearly requires the nominated person to refund to the purchaser the sum that the purchaser paid to the landlord for the surrender, with section 12A(3)(a), which arguably requires the nominated person to pay a second deposit to the landlord but does not require any refund to the purchaser of the deposit that was paid to the landlord. See para 4-334.

### 4-385
12C(5) Where the premises subject to the relevant tenancy included premises other than premises to which this Part applied at the time of the disposal, a notice under this section shall—

(a) require the purchaser only to grant a new tenancy relating to the premises to which this Part then applied, and
(b) require him to do so on the terms referred to in subsection (2) subject to any necessary modifications.

12C(6) The purchase notice may specify the subject-matter of the disposal, and the terms on which the disposal is to be made (whether doing so expressly or by reference to the original disposal), or may provide for those matters to be determined by a leasehold valuation tribunal.

## 4-386

Section 12C(5) and (6) deal with the situation where the surrender by the landlord to the third party purchaser related both to premises within the 1987 Act and to premises outside the Act. In such a case, the tenants can only exercise their rights in relation to those premises that fall within the Act, and the new lease granted to the nominated person is amended as necessary. Either the tenants' notice can specify the subject matter and terms of the new lease, or the notice can provide for this to be determined by a leasehold valuation tribunal.

2It is interesting to contrast this provision with section 5(3), which requires the landlord to sever a transaction that relates to more than one building. There appears to be nothing in the Act to require a landlord to sever a transaction that relates to both premises within the Act and premises outside the Act (it may be that the meaning of "building" precludes the inclusion of premises within the Act and premises outside the Act within a single building—see para 4-17). There also appears to be nothing in sections 11–17 that requires a disposal relating to more than one building to be severed.

## 4-387
### Purchase notice

This is not defined in section 12C. However, section 12B(2) defines the notice served by the tenants on the purchaser under that section as a "purchase notice": see para 4-358. Presumably therefore, "purchase notice" under section 12C(6) means the notice served by the tenants on the purchaser under section 12C(2).

## 4-388

**12D. Nominated persons: supplementary provisions.**

12D(1) The person or persons initially nominated for the purposes of section 12A, 12B or 12C shall be nominated in the notice under that section.

## 4-389

Where the landlord has complied with the obligations under the 1987 Act and served section 5 notices on the qualifying tenants, the requisite majority of qualifying tenants must serve both an acceptance notice and a nomination notice (which may take the form of a single notice). The nomination notice will nominate the nominated person. See para 4-137 for more details.

Where, however, the landlord has failed to comply with the obligations under the 1987 Act, sections 12A, 12B and 12C imply (but do not expressly state) that the notice served on the landlord or the purchaser (as appropriate), whereby the requisite majority of qualifying tenants elect to take the benefit of the contract, or require to take the property disposed of, or require the grant of a new tenancy in place of the tenancy that was surrendered (as the case may be), will also nominate the nominated person. Section 12D(1) confirms that this implication is correct.

## 4-390

> 12D(2) A person nominated for those purposes by the requisite majority of qualifying tenants of the constituent flats may be replaced by another person so nominated if, and only if, he has (for any reason) ceased to be able to act as a nominated person.

## 4-391

The wording of this section is identical to that of section 6(6): see para 4-142.

## 4-392

> 12D(3) Where two or more persons have been nominated and any of them ceases to act without being replaced, the remaining person or persons so nominated may continue to act.

## 4-393

The wording of this section is identical to that of section 6(7): see para 4-144.

## 4-394

> 12D(4) Where, in the exercise of its power to award costs, the court or the Lands Tribunal makes, in connection with any proceedings arising

under or by virtue of this Part, an award of costs against the person or persons so nominated, the liability for those costs is a joint and several liability of that person or those persons together with the qualifying tenants by whom the relevant notice was served.

## 4-395
## Joint and several liability
See para 4-245.

## 4-396
The case of *Kensington Heights Commercial Co Ltd* v *Campden Hill Developments Ltd* concerned a large block of flats in Kensington. The head tenant entered into a contract with the freeholder to surrender the head lease and to take a new lease of a slightly smaller area of land (a boundary strip that had been included in the old head lease was excluded from the new head lease, being of value to the freeholder in connection with a proposed development, but the subtenants (the individual owners of the flats) had no rights over the strip). The surrender was completed and the new lease was granted. The freeholder then agreed to sell the boundary strip to a third party developer. The head tenant then granted a subunderlease of part of the roof to a telecommunications company. No notices under the 1987 Act were served on the individual flat subtenants in relation to any of these transactions, despite the fact that the building fell within the provisions of the Act and there were sufficient qualifying tenants for the Act to apply to a relevant disposal.

The individual flat subtenants served notice on the head tenant under section 12B of the 1987 Act, requiring the head tenant to assign its new head lease (free from the subunderlease to the telecommunications company) to the company nominated by the majority of the subtenants. At first instance, the Court ordered the head tenant to assign the new head lease to the nominated company, but subject to the subunderlease to the telecommunications company.

The Court of Appeal allowed the head tenant's appeal. The "relevant disposal" under the 1987 Act was the agreement to surrender the original head lease, and if the subtenants had served a section 12C notice on the freeholder, they could have required the freeholder to grant a new head lease, in the form of the head lease that was surrendered but for the residue of the term of that head lease, to the nominated company. Neither the agreement to grant the new head lease nor the new head lease itself was a disposal by the head tenant (the immediate landlord of the subtenants), and accordingly the subtenants could not enforce the assignment of that new head lease to their nominated company.

The various disposals in this case can be analysed as follows:

- The agreement to surrender the original head lease was a relevant disposal for the purposes of the 1987 Act. The disposal was made by the head tenant, who was the immediate landlord of the subtenants. Section 12B relates to a disposal which consisted of entering into a contract. However, a section 12B notice should be served on the disponee (ie the freeholder), requiring the freeholder to dispose of the estate or interest that was the subject matter of the original disposal. In the case of a surrender, that estate or interest no longer exists following completion—it has merged in the freehold reversion.

    The correct notice for the subtenants to have served was a section 12C notice, as section 12C applies where the disposal consisted of the surrender of a tenancy, and provides for a new lease to be granted by the landlord reversioner. "Surrender" includes a contract to surrender. The section 12C notice should have been served on the freeholder, being the party to whom the disposal was made, and who was in a position to grant a new lease.

    Note that if the subtenants had correctly served a section 12C notice on the freeholder, their nominated company would have been granted a lease that took effect in reversion on the head tenant's new lease, ie the nominated company would have become the head tenant's landlord, and would have taken subject to the head tenants' new head lease.

    Even if the freeholder had completed the sale of the boundary strip, the subtenants could still have procured the grant of a new lease of the boundary strip. They would have needed to serve a section 12C notice on the third party purchaser developer, as well as a separate 12C notice on the freeholder in relation to the rest of the land the subject of the former head lease.

- The agreement to grant the new head lease was not a relevant disposal for the purposes of the 1987 Act, being made by the freeholder, who at the relevant time was not the immediate landlord of the subtenants. The subtenants could not therefore compel the head tenant to assign the new lease to their nominated company.

- The surrender and the grant of the new head lease were each carried out pursuant to an existing contract, and accordingly neither was a relevant disposal for the purposes of the 1987 Act.

- The grant of the subunderlease to the telecommunications company was a relevant disposal for the purposes of the 1987 Act, being made by the head tenant, the subtenants' immediate landlord, and the subtenants could have procured the assignment of that subunderlease to their nominated company by serving a section 12B notice on the telecommunications company.

Sadly, however, the subtenants served the wrong notice on the wrong party, seeking to take an assignment of an interest to which they were not entitled, and accordingly achieved nothing. The case illustrates the enormous complexity of the 1987 Act and how easy it is to serve the wrong notice or to serve notice on the wrong party.

### 4-397
**13. Determination by rent assessment committees of questions relating to purchase notices.**

13(1) A leasehold valuation tribunal has jurisdiction to hear and determine—

(a) any question arising in relation to any matters specified in a notice under section 12A, 12B or 12C, and
(b) any question arising for determination as mentioned in section 8C(4), 12A(5) or 12B(4) (matters left for determination by tribunal).

### 4-398
### Notice under section 12A, 12B or 12C
This is a reference to the notice served on the landlord or the purchaser (as appropriate), whereby the requisite majority of qualifying tenants elect to take the benefit of the contract, or require to take the property disposed of, or require the grant of a new tenancy in place of the tenancy that was surrendered (as the case may be).

## 4-399
## Any question arising for determination as mentioned in sections 8C(4), 12A(5) or 12B(4)

Section 8C(4) is the anti-avoidance provision that deals with a disposal wholly or partly for non-monetary consideration. The valuation of the non-monetary consideration (which is to be calculated as its value in the hands of the landlord) may be determined by the Leasehold Valuation Tribunal.

Sections 12A(5) and 12B(4) deal with the situation where the original disposal related both to property within the 1987 Act and to property outside the Act. The Leasehold Valuation Tribunal may be asked to determine what should be transferred to the nominated person and/or the terms on which the transfer should take place.

## 4-400

13(2) On an application under this section the interests of the persons by whom the notice was served under section 12A, 12B or 12C shall be represented by the nominated person; and accordingly the parties to any such application shall not include those persons.

## 4-401

It is clear from section 13(2) that the tenants may not be a party to any application to the Leasehold Valuation Tribunal under sections 12A/12B/12C, despite the fact that it is the tenants' notice that forms the subject matter of the application to the Tribunal; their interests will be represented solely by the nominated person. The section does not refer to the determination of matters under section 8C(4); section 8C(4) expressly states that the landlord or the nominated person may apply to the Tribunal, but the tenants are not specifically excluded from making such an application.

## 4-402

**14. Withdrawal of nominated person from transaction.**

14(1) Where notice has been duly served on the landlord under—

section 12B (right of qualifying tenants to compel sale, &c by purchaser), or
section 12C (right of qualifying tenants to compel grant of new tenancy by superior landlord),

the nominated person may at any time before a binding contract is entered into in pursuance of the notice, serve notice under this section on the purchaser (a "notice of withdrawal") indicating an intention no longer to proceed with the disposal.

## 4-403
Section 14(1) gives the nominated person, after a notice has been served under section 12B or section 12C, the right (but does not impose any obligation) to serve notice on the third party purchaser withdrawing from the transaction (defined as a "notice of withdrawal"). Obviously, this right can only be exercised before a binding contract is entered into.

## 4-404
### Duly served on the landlord
Sections 12B and 12C in fact require the requisite majority's notice to be served on the third party purchaser rather than on the landlord. Presumably this is a simple drafting mistake.

## 4-405
### Nominated person
See para 4-130.

## 4-406
### Serve
See para 4-82.

## 4-407
### Notice
See para 4-83.

## 4-408
### Purchaser
See para 4-312. See also para 4-458.

## 4-409
14(2) If at any such time the nominated person becomes aware that the number of qualifying tenants of the constituent flats desiring to

proceed with the disposal is less than the requisite majority of those tenants, he shall forthwith serve a notice of withdrawal.

## 4-410
Where section 14(1) is optional, section 14(2) is mandatory. It obliges the nominated person to serve a notice of withdrawal as soon as the nominated person becomes aware that the number of supporting tenants has fallen below the requisite majority. See also para 4-130, in relation to the tenants' right to withdraw their support.

## 4-411
### Requisite majority of qualifying tenants
See para 4-99.

## 4-412
### Notice of withdrawal (nominated person)
See para 4-403.

## 4-413
14(3) If a notice of withdrawal is served under this section the purchaser may recover from the nominated person any costs reasonably incurred by him in connection with the disposal down to the time when the notice is served on him.

## 4-414
Unlike in the case of a notice of withdrawal under section 9A, where early withdrawal by the nominated person brings with it no costs liability, under section 14(3), the nominated person is liable for the costs of the third party purchaser regardless of the timing of service of notice of withdrawal. Additionally, the nominated person alone is liable for the purchaser's costs, and there is no shared liability with the tenants.

## 4-415
### In connection with the disposal
Although it would seem to be logical for the purchaser only to be able to recover costs in so far as they relate to the proposed disposal to the nominated person, these words imply that the purchaser can in fact recover costs, provided that they are reasonably incurred, in

## 4-416

    14(4)    If a notice of withdrawal is served at a time when proceedings arising under or by virtue of this Part are pending before the court or the Lands Tribunal, the liability of the nominated person for any costs incurred by the purchaser as mentioned in subsection (3) shall be such as may be determined by the court or (as the case may be) by the Tribunal.

    14(5)    The costs that may be recovered by the purchaser under this section do not include any costs incurred by him in connection with an application to a leasehold valuation tribunal.

## 4-417

Section 14(4) provides that, where there are proceedings pending before the court or Lands Tribunal, the nominated person's liability for the third party purchaser's costs under section 14(3) will be determined by the court or Tribunal. Section 14(5) specifically excludes any costs incurred by the purchaser in connection with an application to a Leasehold Valuation Tribunal.

## 4-418

    **16.**    **Right of qualifying tenants to compel sale etc. by subsequent purchaser.**

    16(1)    This section applies where, at the time when a notice is served on the purchaser under section 11A, 12A, 12B or 12C, he no longer holds the estate or interest that was the subject-matter of the original disposal.

## 4-419

Note that there is now no section 15 in the 1987 Act!

Section 16(1) states that section 16 applies where the original third party purchaser has already parted with the property that was the subject of the original disposal by the time the tenants serve notice upon the original third party purchaser. Section 16(2) deals with the obligations on the original purchaser where the notice served was under section 11A, and section 16(3) deals with the original purchaser's obligations where a section 12A/B/C notice was served. Basically, in every case, the original purchaser must forward a copy of the tenants' notice to the subsequent purchaser and give the tenants details of that subsequent purchaser. Once the original purchaser has complied with obligations in sections 16(2) or 16(3) as appropriate, the original purchaser's obligations in relation to the property are at an end and become obligations of the subsequent purchaser.

## 4-420
## Purchaser
See para 4-312. See also para 4-458.

## 4-421
## Original disposal
See para 4-305.

## 4-422
16(2)  In the case of a notice under section 11A (right to information as to terms of disposal, &c.) the purchaser shall, within the period for complying with that notice—

(a) serve notice on the person specified in the notice as the person to whom particulars are to be provided of the name and address of the person to whom he has disposed of that estate or interest ( 'the subsequent purchaser'), and

(b) serve on the subsequent purchaser a copy of the notice under section 11A and of the particulars given by him in response to it.

## 4-423
## Within the period for complying with that notice
Section 11A(4) requires the third party purchaser to comply with the tenants' notice seeking information within one month of the date of service of that notice.

## 4-424
## Serve
See para 4-82.

## 4-425
## Notice
See para 4-83.

## 4-426
## Person specified in the notice
This is the person referred to in section 11A(2) as being the person to whom information is to be supplied on behalf of the tenants.

## 4-427
## Subsequent purchaser
Section 16(2)(a) defines this as the person to whom the original third party purchaser has disposed of that estate or interest transferred by the landlord in breach of the 1987 Act. Note that references in section 17 to the purchaser include the subsequent purchaser: section 17(7), para 4-458.

## 4-428
16(3)  In the case of a notice under section 12A, 12B or 12C the purchaser shall forthwith—

 (a) forward the notice to the subsequent purchaser, and
 (b) serve on the nominated person notice of the name and address of the subsequent purchaser.

## 4-429
## Notice under section 12A, 12B or 12C
This is a reference to the notice served on the landlord or the purchaser (as appropriate), whereby the requisite majority of qualifying tenants elect to take the benefit of the contract, or require to take the property disposed of, or require the grant of a new tenancy in place of the tenancy that was surrendered (as the case may be).

## 4-430
## Purchaser
Note that in fact, notice under section 12A must be served by the tenants on the landlord, not on the original third party purchaser. This appears to have been overlooked in the drafting of section 16(3).

## 4-431
## Nominated person
See para 4-130.

## 4-432
16(4)  Once the purchaser serves a notice in accordance with subsection (2)(a) or (3)(b), sections 12A to 14 shall, instead of applying to the purchaser, apply to the subsequent purchaser as if he were the transferee under the original disposal.

## 4-433
## Sections 12A to 14
Once the original third party purchaser has forwarded the tenants' notice to the subsequent purchaser and has informed the tenants of this, any reference to the original third party purchaser in sections 12A to 14 of the 1987 Act shall apply instead to the subsequent purchaser. However:

- Section 11A (obligation to give the tenants information about the disposal) is not referred to here. This obligation therefore remains with the original third party purchaser and is not passed to the subsequent purchaser.

- Section 12A in fact involves the tenants serving notice on the landlord, not the third party purchaser.

## 4-434
16(5)  Subsections (1) to (4) have effect, with any necessary modifications, in a case where, instead of disposing of the whole of the estate or interest referred to in subsection (1) to another person, the purchaser has disposed of it in part or in parts to one or more other persons.

In such a case, sections 12A to 14—

(a) apply to the purchaser in relation to any part of that estate or interest retained by him, and
(b) in relation to any part of that estate or interest disposed of to any other person, apply to that other person instead as if he were (as respects that part) the transferee under the original disposal.

## 4-435

Where part only of the property disposed of to the original third party purchaser has been passed on to the subsequent purchaser (the original purchaser retaining the rest of it), or where the original purchaser has divided the property between more than one subsequent purchaser, the provisions of section 16(1)–(4) apply separately to each part and sections 12A to 14 apply to each such purchaser in relation to the part of the property held by that purchaser at the relevant time.

## 4-436

**17. Termination of rights against new landlord or subsequent purchaser.**

17(1) If, at any time after a notice has been served under section 11A, 12A, 12B or 12C, the premises affected by the original disposal cease to be premises to which this Part applies, the purchaser may serve a notice on the qualifying tenants of the constituent flats stating—

(a) that the premises have ceased to be premises to which this Part applies, and
(b) that any such notice served on him, and anything done in pursuance of it, is to be treated as not having been served or done.

## 4-437

As previously stated (see para 4-178), the landlord is never *obliged* to serve a notice withdrawing from the transaction. Section 10(1) makes it clear that this principle still applies even where the 1987 Act ceases to apply to the premises after the service of the offer notice but before contracts for the disposal have been exchanged. The landlord may choose to serve a notice under this section on the qualifying tenants, and thereupon the Act shall cease to apply to the proposed disposal — if, however, the landlord chooses not to serve such a notice, then the Act will continue to apply to the proposed disposal as provided by section 10(3).

Section 17 contains equivalent provisions in the case of the purchaser to those that apply to the landlord in section 10.

## 4-438
## Notice ... served under section 11A
This is the tenants' notice requesting information about the disposal under section 11A.

## 4-439
## Notice ... served under section ... 12A, 12B or 12C
See para 4-398.

## 4-440
## Original disposal
See para 4-305.

## 4-441
## Purchaser
See para 4-312. See also para 4-458.

## 4-442
## Serve
See para 4-82. See also para 4-273.

## 4-443
## Notice
See para 4-83.

## 4-444
## Qualifying tenants
See para 4-37.

## 4-445
## Constituent flats (section 11)
See para 4-309.

## 4-446
    17(2)    A landlord who has not served such a notice on all of the qualifying tenants of the constituent flats shall nevertheless be treated as having duly served a notice under subsection (1)—

Section by Section

(a) if he has served such a notice on not less than 90% of those tenants, or
(b) where those qualifying tenants number less than ten, if he has served such a notice on all but one of them.

## 4-447

Section 17(2) (like section 10(2), see para 4-273) provides that, even where the landlord omits to serve notice on one qualifying tenant (where there are fewer than 10 qualifying tenants) or omits to serve notices on 10% of the qualifying tenants (where there are 10 or more qualifying tenants), the landlord will nevertheless be treated as having served notice pursuant to section 17(1).

## 4-448

17(3) Where a period of three months beginning with the date of service of a notice under section 12A, 12B or 12C on the purchaser has expired—

(a) without any binding contract having been entered into between the purchaser and the nominated person, and
(b) without there having been made any application in connection with the notice to the court or to a leasehold valuation tribunal,

the purchaser may serve on the nominated person a notice stating that the notice, and anything done in pursuance of it, is to be treated as not having been served or done.

## 4-449

Section 17(3) effectively gives the nominated person three months from the date on which the nominated person starts to enforce the tenants' rights against the purchaser within which to either exchange contracts or apply to the court or Leasehold Valuation Tribunal. If at the end of that three-month period, the nominated person has done neither, then the purchaser may bring the whole thing to an end by serving notice on the nominated person, nullifying anything done so far. The purchaser's notice will terminate the tenants' rights against the purchaser and there is no possibility of serving a second tenants' notice: see section 17(5), para 4-454.

It is curious that there is no express obligation on the purchaser to accept an exchange of contracts offered by the nominated person.

Section 8A(6) stops short of obliging the landlord to accept such an exchange of contracts, which is logical because it is always open to the landlord to decide not to proceed with the disposal at any time. In the case of the purchaser, however, it is crucial, from the tenants' point of view, that they can compel the purchaser to transfer the purchaser's rights or interest to the nominated person. Section 17(3)(b), however, implies (but does not expressly state) that failure on the part of the purchaser to accept an exchange of contracts will entitle the nominated person to apply to court for relief. Presumably it is intended that service of notice under section 19(2), requiring the default of the purchaser in failing to accept an exchange of contracts to be made good, should be followed by an application to the court by the nominated person.

Note also that this section appears to contemplate exchange of a further contract between the nominated person and purchaser where section 12A applies, despite the fact that section 12A itself provides that by serving notice under that section, the nominated person will step into the shoes of the purchaser under the original contract, so that there seems to be no need for a second contract to be exchanged.

## 4-450
### Court

Section 60(1) of the 1987 Act defines "court" to mean either county court or High Court. Section 52 provides that a county court shall have jurisdiction to hear and determine any question arising under any provision of the Act that relates to tenants' rights of first refusal, except for a question falling within the jurisdiction of a leasehold valuation tribunal, as well as any other proceedings joined with those proceedings. A person who commences proceedings in the High Court may be limited to recovery of costs on county court scales, except where the proceedings were brought in the High Court to enable them to be joined with existing proceedings.

## 4-451

17(4) Where any such application as is mentioned in subsection (3)(b) was made within the period of three months referred to in that subsection, but—

  (a) a period of two months beginning with the date of the determination of that application has expired,
  (b) no binding contract has been entered into between the purchaser and the nominated person, and

(c) no other such application as is mentioned in subsection (3)(b) is pending,

the purchaser may serve on the nominated person a notice stating that any notice served on him under section 12A, 12B or 12C, and anything done in pursuance of any such notice, is to be treated as not having been served or done.

## 4-452
Section 17(4) is really an extension to, and continuation of, section 17(3). As previously stated, the nominated person is safe from the purchaser's notice, nullifying the tenants' rights, provided that the nominated person either exchanges contracts within the three-month period or applies to the court or tribunal. Section 17(4) then deals with the situation where an application has been made but, following the determination of that application, the nominated person has failed, for a period of two months, either to exchange contracts or to make some other court or tribunal application (perhaps an appeal against the initial determination). In that case, the purchaser may bring the whole thing to an end by serving notice on the nominated person, nullifying anything done so far. The purchaser's notice will terminate the tenants' rights against the purchaser: see section 17(5), para 4-454 and the case of *Boyle v Hallstate Ltd*.

## 4-453
17(5) Where the purchaser serves a notice in accordance with subsection (1), (3) or (4), this Part shall cease to have effect in relation to him in connection with the original disposal.

## 4-454
As previously stated, service by the purchaser of notice under section 17(1) (where the premises are no longer within the 1987 Act), section 17(3) (where the nominated person has failed to exchange or make an application within three months) or section 17(4) (where the nominated person has failed to exchange or make a further application within two months) will terminate the tenants' rights against the purchaser in connection with the unauthorised disposal: see *Boyle v Hallstate Ltd*.

## 4-455
### In relation to him in connection with the original disposal
Note that the purchaser's notice does not take the premises outside the 1987 Act, it merely terminates the tenant's rights to acquire the purchaser's interest derived from the original unauthorised disposal.

## 4-456
    17(6)    Where a purchaser is entitled to serve a notice under subsection (1) but does not do so, this Part shall continue to have effect in relation to him in connection with the original disposal as if the premises in question were still premises to which this Part applies.

## 4-457
As in the case of section 10(1) (para 4-267), service of notice under section 17(1) is entirely optional. If the purchaser chooses not to serve such a notice, then the Act will continue to apply to the unauthorised disposal as provided by this section. Although at first glance it may seem odd that premises that fall outside the 1987 Act are nonetheless deemed still to fall within the Act simply because the purchaser has not bothered to serve a notice under section 17(1), this may be justified by the fact that the original disposal was unauthorised and made in respect of premises which, at the time of that original disposal, did fall within the Act. On the other hand, it seems unfair, from the tenants' perspective, that they lose their rights in relation to the unauthorised disposal simply because the premises have subsequently fallen outside the Act.

## 4-458
    17(7)    References in this section to the purchaser include a subsequent purchaser to whom sections 12A to 14 apply by virtue of section 16(4) or (5).

## 4-459
Section 17(7) confirms the concept that premises that are the subject of an unauthorised disposal remain "tainted" by that disposal unless and until the tenants are notified of the disposal and their rights in relation to it, and the time period for exercising those rights expires. A subsequent disposal by the original purchaser to a subsequent purchaser does not affect the position.

## 4-460
**18. Notices served by prospective purchasers to ensure that rights of first refusal do not arise.**

18(1) Where—

(a) any disposal of an estate or interest in any premises consisting of the whole or part of a building is proposed to be made by a landlord, and
(b) it appears to the person who would be the transferee under that disposal ('the purchaser') that any such disposal would, or might, be a relevant disposal affecting premises to which this Part applies,

the purchaser may serve notices under this subsection on the tenants of the flats contained in the premises referred to in paragraph (a) ('the flats affected').

## 4-461

As previously stated, the position of the purchaser, in the case of an unauthorised disposal, is actually more precarious than that of the landlord, inasmuch as that disposal will be valid (so the landlord will have successfully disposed of the premises, despite the fact that the landlord may be guilty of a criminal offence) but the tenants then enjoy significant rights against the purchaser, who may lose the premises. For this reason, it is crucial that the purchaser is able to ensure that the tenants will not have rights against the purchaser following completion of the purchase.

A prospective purchaser may, of course, be prepared to rely on the section 5 notices served by the landlord, and may wish to obtain copies of these and check that everything has been done properly. Alternatively, the prospective purchaser may prefer to use section 18 to check directly as to the position.

Section 18 enables the prospective purchaser to serve notices that are similar to those served by a landlord under section 5. However, the prospective purchaser does not necessarily know which of the tenants are qualifying tenants (unlike the landlord, the prospective purchaser has no direct contact with the tenants), and therefore serves notices on all the residential tenants. The notices ask for three pieces of information:

- Whether the tenant has received a section 5 notice.

- If not, whether the tenant knows of any reason why the tenant is not entitled to a section 5 notice (perhaps because the tenant is an assured shorthold tenant).

- If the tenant knows of no such reason, whether the tenant would respond positively to a section 5 notice if it were to be served.

If a majority of the tenants respond positively to the second question (ie they know of a reason why they are not entitled to a section 5 notice) or negatively to the third question (ie they would not respond positively to a section 5 notice if one were to be served on them) then the purchaser can proceed with the disposal, secure in the knowledge that the premises fall outside the Act.

### 4-462
### Disposal of an estate or interest in any premises
The 1987 Act applies to a disposal only where it is a "relevant disposal"—see para 4-47. Section 18(1) entitles the prospective purchaser to serve notices where the prospective purchaser believes that the proposed disposal will in fact constitute a relevant disposal: section 18(1)(b).

### 4-463
### Building
See para 4-17.

### 4-464
### Proposed to be made
See para 4-79.

### 4-465
### Landlord
See para 4-29.

## 4-466
### Purchaser (section 18)
"Purchaser" is defined in section 18(1)(b) as the person who would be the transferee under the proposed disposal if it were to proceed. Note that this is not the same "purchaser" as the person defined as such in section 11(3), see para 4-312.

## 4-467
### Relevant disposal
See para 4-47.

## 4-468
### Affecting premises
See para 4-04.

## 4-469
### Premises
See para 4-05.

## 4-470
### Serve
See para 4-82.

## 4-471
### Notice
See para 4-83.

## 4-472
### Tenants
See para 4-36.

## 4-473
### Flats
See para 4-19.

## 4-474

18(2) Any notice under subsection (1) shall—

(a) inform the person on whom it is served of the general nature of the principal terms of the proposed disposal, including in particular—

(i) the property to which it would relate and the estate or interest in that property proposed to be disposed of by the landlord, and

(ii) the consideration required by him for making the disposal;

(b) invite that person to serve a notice on the purchaser stating—

(i) whether the landlord has served on him, or on any predecessor in title of his, a notice under section 5 with respect to the disposal, and

(ii) if the landlord has not so served any such notice, whether he is aware of any reason why he is not entitled to be served with any such notice by the landlord, and

(iii) if he is not so aware, whether he would wish to avail himself of the right of first refusal conferred by any such notice if it were served; and

(c) inform that person of the effect of the following provisions of this section.

## 4-475
### Principal terms

See para 4-97 (section 5A) or para 4-105 (section 5B) or para 4-112 (section 5C) or para 4-117 (section 5D) (as applicable). Section 18(2) refers to the "general nature" of the principal terms, which implies that the prospective purchaser in the section 18 notice can be less precise as to the details of the proposed disposal than the landlord in the section 5A/5B/5C/5D notice.

## 4-476

18(3) Where the purchaser has served notices under subsection (1) on at least 80 per cent. of the tenants of the flats affected and—

(a) not more than 50 per cent. of the tenants on whom those notices have been served by the purchaser have served notices on him in pursuance of subsection (2)(b) by the end of the period of two months beginning with the date on which the last of them was served by him with a notice under this section, or

(b) more than 50 per cent. of the tenants on whom those notices have been served by the purchaser have served notices on him in pursuance of subsection (2)(b) but the notices in each case indicate that the tenant serving it either—
  (i) does not regard himself as being entitled to be served by the landlord with a notice under section 5 with respect to the disposal, or
  (ii) would not wish to avail himself of the right of first refusal conferred by such a notice if it were served,

the premises affected by the disposal shall, in relation to the disposal, be treated for the purposes of this Part as premises to which this Part does not apply.

## 4-477
## On at least 80% of the tenants of the flats affected

The prospective purchaser only needs to serve section 18 notices on 80% of the tenants of the "flats affected" (defined in section 18(1) as being all the flats in the relevant premises). There is no guidance as to which tenant(s) should or should not receive notices—it is entirely up to the prospective purchaser whether to serve all the tenants or choose to exclude up to 20% of them (and which 20% to exclude).

More than 50% of those tenants on whom notices have been served must then respond positively in order to preserve their rights. If, therefore, the prospective purchaser has only served notices on 80% of the tenants, then a positive response is only required from more than 40% of the total number of tenants. A minority of the tenants in the building will be sufficient, in this situation, to preserve the rights of first refusal.

This contrasts with the position in relation to the landlord's notices under section 5, where the landlord must serve at least 90% of the qualifying tenants (where there are more than 10 flats) and where more than 50% of the total number of qualifying tenants must respond positively in order to exercise the tenants' rights.

## 4-478

18(4) For the purposes of subsection (3) each of the flats affected shall be regarded as having one tenant, who shall count towards any of the percentages specified in that subsection whether he is a qualifying tenant of the flat or not.

## 4-479

Each flat occupied by a qualifying tenant is deemed to have a single vote for this purpose. The Act does not deal with what would happen if a tenant of a particular flat in fact comprised more than one individual, and each individual voted differently.

## 4-480

18A. **The requisite majority of qualifying tenants.**

18A(1) In this Part 'the requisite majority of qualifying tenants of the constituent flats' means qualifying tenants of constituent flats with more than 50 per cent of the available votes.

18A(2) The total number of available votes shall be determined as follows—

(a) where an offer notice has been served under section 5, that number is equal to the total number of constituent flats let to qualifying tenants on the date when the period specified in that notice as the period for accepting the offer expires;

(b) where a notice is served under section 11A without a notice having been previously served under section 5, that number is equal to the total number of constituent flats let to qualifying tenants on the date of service of the notice under section 11A;

(c) where a notice is served under section 12A, 12B or 12C without a notice having been previously served under section 5 or section 11A, that number is equal to the total number of constituent flats let to qualifying tenants on the date of service of the notice under section 12A, 12B or 12C, as the case may be.

18A(3) There is one available vote in respect of each of the flats so let on the date referred to in the relevant paragraph of subsection (2), which shall be attributed to the qualifying tenant to whom it is let.

18A(4) The persons constituting the requisite majority of qualifying tenants for one purpose may be different from the persons constituting such a majority for another purpose.

## 4-481
See para 4-99.

## 4-482
**19. Enforcement of obligations under Part I.**

19(1) The court may, on the application of any person interested, make an order requiring any person who has made default in complying with any duty imposed on him by any provision of this Part to make good the default within such time as is specified in the order.

## 4-483
Section 19(1) provides a means for enforcing any duty imposed by the 1987 Act by application to the court.

## 4-484
### Any person interested
"Interested person" has been defined by case law to mean a person having a pecuniary or proprietary right in, or likely to be biased by, the action complained of: *Bearmans Ltd* v *Metropolitan Police Receiver* [1961] 1 WLR 634.

## 4-485
19(2) An application shall not be made under subsection (1) unless

(a) a notice has been previously served on the person in question requiring him to make good the default, and
(b) more than 14 days have elapsed since the date of service of that notice without his having done so.

## 4-486
Section 19(2) ensures that no person having a duty under the Act can be taken to court until they have been notified of their default and given at least 14 days within which to rectify this.

## 4-487

19(3) The restriction imposed by section 1(1) may be enforced by an injunction granted by the court.

## 4-488
**20. Construction of Part I and power of Secretary of State to prescribe modifications.**

20(1) In this Part—
'acceptance notice' has the meaning given by section 6(3);

See para 4-133.

'associated company', in relation to a body corporate, means another body corporate which is (within the meaning of section 736 of the Companies Act 1985) that body's holding company, a subsidiary of that body or another subsidiary of that body's holding company; ...

See para 4-62.

'constituent flat' shall be construed in accordance with section 5(1) or 11(2), as the case may require; ...

See para 4-100 (section 5) and para 4-309 (section 11).

'disposal' shall be construed in accordance with section 4(3) and section 4A (application of provisions to contracts) and references to the acquisition of an estate or interest shall be construed accordingly; ...

The 1987 Act does not apply to every disposal, but only to a "relevant disposal"—see para 4-47.

'landlord', in relation to any premises, shall be construed in accordance with section 2; ...

See para 4-29.

'the nominated person' means the person or persons for the time being nominated by the requisite majority of the qualifying tenants of the constituent flats; ...

See para 4-130.

'offer notice' means a notice served by a landlord under section 5; ...

See para 4-125.

'the original disposal' means the relevant disposal referred to in section 11(1); ...

See para 4-305.

'the protected interest' means the estate, interest or other subject-matter of an offer notice; ...

See para 4-129.

'the protected period' has the meaning given by section 6(4); ...

See para 4-135.

'purchase notice' has the meaning given by section 12B(2); ...

See para 4-358.

'purchaser' has the meaning given by section 11(3); ...

See para 4-312. Note that "purchaser" is separately defined in section 18(1), see para 4-458.

'qualifying tenant', in relation to a flat, shall be construed in accordance with section 3; ...

See para 4-37.

'relevant disposal' shall be construed in accordance with section 4; ...

See para 4-47.

'the requisite majority', in relation to qualifying tenants, shall be construed in accordance with section 18A; ...

See para 4-99.

'transferee', in relation to a disposal, shall be construed in accordance with section 4(3).

The term "transferee" is clearly intended under the 1987 Act to mean the person who takes an interest in land, even where it would be natural to describe that person differently, such as in the case of the surrender of a lease, where the person receiving the surrender would more naturally be described as the landlord rather than the transferee.

**4-489**
    (2)    In this Part—

- (a) any reference to an offer is a reference to an offer made subject to contract, and
- (b) any reference to the acceptance of an offer is a reference to its acceptance subject to contract.

Section 20(2) simply confirms that, by serving or accepting an offer notice, no contract is created, because both the offer and the acceptance are made "subject to contract".

**4-490**
    (3)    Any reference in this Part to a tenant of a particular description shall be construed, in relation to any time when the interest under his tenancy has ceased to be vested in him, as a reference to the person who is for the time being the successor in title to that interest.

It is not unusual for the tenant of an individual flat to change during the process of acquisition of the landlord's interest under the 1987 Act. Section 20(3) makes it clear that the person having rights under the Act is the person who is from time to time the tenant of the flat, however the interest may have become vested in that person. This may affect the total number of tenants exercising—or failing to exercise—their rights under the Act.

See para 4-36 in relation to the meaning of "tenant"—the Act itself does not define the meaning of this word. However, normally a tenant will be the owner of the legal estate in the lease: see the case of *Brown & Root Technology Ltd* v *Sun Alliance and London Assurance Co Ltd* [1996] EWCA Civ 1261, which held that where a lease had been transferred from one party to another but the transferee had not registered the transfer at H M Land Registry, the transferor remained "the tenant" under the lease.

**4-491**
    (4)    The Secretary of State may by regulations make such modifications of any of the provisions of sections 5 to 18 as he considers appropriate, and any such regulations may contain such incidental, supplemental or transitional provisions as he considers appropriate in connection with the regulations.

    (5)    In subsection (4) 'modifications' includes additions, omissions and alterations.

Any order or regulation made by the Secretary of State under the 1987 Act shall be exercisable by statutory instrument: section 53(1) of the Act. A statutory instrument that amends the percentages applicable to mixed use properties under section 1(5) of the Act will be subject to annulment pursuant to a resolution of either House of Parliament: section 53(2) of the Act.

# Collective Enfranchisement Rights and Rights to Manage

## A. The Leasehold Reform, Housing and Urban Development Act 1993 (as amended) —collective enfranchisement (flats)

5-01
Subject to meeting various qualifying criteria, residential tenants of flats in either purely residential buildings or mixed commercial/residential buildings have the right under the Leasehold Reform, Housing and Urban Development Act 1993 to buy the freehold of their building from the freeholder. These are freestanding rights and do not depend on the freeholder proposing to dispose of an interest in the building, unlike the rights granted by the Landlord and Tenant Act 1987 (as amended).

5-02
The qualifying conditions as originally drafted caused particular difficulties for those seeking to exercise their rights. Amendments introduced under the Commonhold and Leasehold Reform Act 2002 have widened the number of properties which fall within the 1993 Act and made it easier for tenants to exercise their rights. Unless indicated otherwise, any changes were brought in on 30 September 2003.

## 5-03
In order to exercise their rights under the 1993 Act, a sufficient number of the qualifying tenants must act together and serve a notice to enfranchise on the freeholder. There is a strict timetable to be followed, breach of which may lead to a deemed withdrawal of the notice and a 12-month prohibition on serving a fresh notice to enfranchise.

## 5-04
Although the tenants must pay the freeholder (and any intermediate landlords) a premium to compensate for the loss of the freehold interest in the building, this may not adequately compensate the freeholder of mixed use premises or of a mixed use development. The strict timetable may deprive the freeholder of the right to object to the tenants' notice to enfranchise or to claim a lease back of non-qualifying units such as commercial parts of the building.

# Legislation

## 5-05
The rights of the tenants are contained in:

- sections 1–38 of the Leasehold Reform, Housing an Urban Development Act 1993, as amended by the Commonhold and Leasehold Reform Act 2002, sets out the main provisions of the right of collective enfranchisement

- sections 90–103 of the 1993 Act contain other key provisions, including anti-avoidance provisions and provisions as to notices

- Schedules 1–10 of the 1993 Act give further detail as to the exercise of the rights

- the Leasehold Reform (Collective Enfranchisement and Lease Renewal) Regulations 1993 (SI 1993/2407) set out some of the conveyancing and other procedures to be followed

- the Leasehold Reform (Collective Enfranchisement) (Counternotices) (England) Regulations 2002 (SI 2002/3208) deal with additional requirements for a landlord's counternotice.

## Qualification

**5-06**

For tenants to exercise their rights under the 1993 Act, the premises and the tenants must meet the qualifying criteria and a sufficient number of qualifying tenants must act together.

**5-07**

Briefly, the Act applies where all the following apply:

1. The premises:

    (a) consist of a self-contained building or part of a building
    (b) contain at least two flats held by qualifying tenants
    (c) are not premises excluded from the provisions of the Act
    (d) contain no more than 25%, by internal floor area, used for non-residential purposes.

2. At least two-thirds of the flats in the premises are let to qualifying tenants.

**5-08**

Where the Act applies, at least 50% of the total number of tenants in the building must then act together to exercise those rights (and all of the 50% must be qualifying tenants).

## Premises

**5-09**

Sections 3 and 4 contain most of the provisions that set out which premises fall within the 1993 Act.

**5-10**

The Act applies to premises which consist of a self-contained building (that is, one that is structurally detached) or a vertically divisible part of a building (that is, a part that could be divided vertically from the rest of the building and redeveloped separately without significant interruption to the services provided to the remainder of the building).

## 5-11

Unfortunately the Act itself contains no definition of "building", but case law assists in certain areas. The inability to separate services shared with a neighbouring building means that the premises do not qualify under the Act: see *Oakwood Court (Holland Park) Ltd* v *Daejan Properties Ltd* [2007] 1 EGLR 121. Where the part of the building is divided both vertically and horizontally from the rest of the building, the part will not be self-contained for the purposes of the Act: see *London Rent Assessment Panel* v *Holding and Management (Solitaire) Ltd* [2007] EW Lands LRX/138/2006.

## 5-12

As previously stated, the Act applies to mixed use premises as well as to purely residential premises, so long as no more than 25% of the internal floor area of the building is for non-residential use. Where a commercial lease permits part of the premises to be used for residential purposes, then that part falls within the residential percentage for the purposes of the Act: see *Gaingold Ltd* v *WHRA RTM Co Ltd* [2006] 1 EGLR 81, where a restaurant lease permitted part of the premises to be used for staff accommodation, and that part was "residential" for the purposes of the Act.

## 5-13

In calculating the ratio between the residential and non-residential parts of a building, the common parts of the building are to be excluded: see *Marine Court (St Leonards on Sea) Freeholders Ltd* v *Rother District Investments Ltd* [2008] 1 EGLR 39.

## 5-14

The building must also contain at least two flats (defined in section 101(1) of the 1993 Act as a separate set of premises constructed or adapted as a dwelling and lying above or below another set of premises). At least two-thirds of the flats must be let to qualifying tenants — see para 5-16–5-20 below.

## 5-15

The following premises are excluded from the provisions of the 1993 Act:

- A building containing four or fewer flats in a non-purpose built block, where the landlord has owned the block since before conversion *and* either the landlord or a member of the landlord's family lives there and has done so for at least 12 months.

- Where the freehold includes operational railway track. This exclusion deals, for example, with the case of a block of flats built over a railway bridge that might need repairing at some stage.

- A building owned by the Crown.

- A building owned by the National Trust (where an interest in the property is vested inalienably in it).

## Qualifying tenant
### 5-16
A qualifying tenant must have a long lease of a flat. "Long lease" is defined in section 7 of the 1993 Act, and generally means an original term of more than 21 years (or a shorter lease which contains a perpetual renewal clause, or which contains an obligation to renew without payment and where it has in fact been renewed to a total of more than 21 years). Leases which have come to an end but are being statutorily continued under Part I of the Landlord and Tenant Act 1954 or under Schedule 10 to the Local Government and Housing Act 1989 may also be included within this definition. Break clauses and forfeiture provisions are disregarded in ascertaining the length of the term.

### 5-17
Where the flat is held on two or more leases with the same landlord and the same tenant, they are treated as one lease and fall within the Act, except where one of those leases is not a long lease.

### 5-18
The following types of lease are excluded:

- Shared ownership leases where the tenant's total share is less than 100%.

- Business leases (defined in section 101 of the 1993 Act as "a tenancy to which Part II of the Landlord and Tenant Act 1954 applies").

- Subleases granted in breach of the superior lease and where the landlord has not subsequently waived that breach.

- Leases granted by a charitable housing trust where the flat forms part of the accommodation provided by the trust in pursuit of its charitable purposes: see *Brick Farm Management Ltd* v *Richmond Housing Partnership Ltd* [2005] EWHC 1650; [2005] 3 EGLR 57.

- A lease held by a tenant who holds three or more leases of flats in the building (excluding any leases that are otherwise excluded under the above points).

## 5-19
Where both the tenant and the subtenant hold long leases, the subtenant will be the qualifying tenant, rather than the tenant. No flat may have more than one qualifying tenant, and joint owners are treated together as a single tenant.

## 5-20
There is no longer a residence requirement, so companies are now eligible to take part in the enfranchisement process.

## Practical considerations for the tenants

### 5-21
Once the enfranchisement process has commenced, a strict timetable comes into effect and failure to comply with the timetable may mean that the tenants are then deemed to have withdrawn from the process (but must pay the landlord's costs) and are also prohibited from making a fresh claim to enfranchise for 12 months. It is therefore important that they not only ascertain whether the building qualifies and contains sufficient qualifying tenants, but also decide how they will fund the acquisition (both the premium and their own and the landlord's costs, and how such sums will be apportioned) and how the freehold of the building will be held.

### 5-22
Although the 1993 Act has been amended to provide that a right to enfranchise company be set up to exercise the enfranchisement right,

*Collective Enfranchisement Rights*

## Procedure

**5-23**

The right to enfranchise is exercised by qualifying tenants comprising more than 50% of the total number of tenants in the building together serving an initial notice on the freeholder to acquire the freehold in the name of a nominee purchaser at a price specified by the tenants. The freeholder must then serve a counternotice accepting or rejecting the claim or setting out alternative proposals (and seeking a leaseback of non-qualifying units, if so required) within (usually) two months.

## Timetable

**5-24**

Once the initial notice has been served, a strict timetable is triggered (see Flowchart F, para 5-60). The consequences of default are generally of most concern to the tenants, and in particular the combined effect, where the nominee purchaser fails to comply with any step in the timetable, of deemed withdrawal of the initial notice and the consequent 12 months' prohibition on serving a fresh initial notice.

- The freeholder may serve a notice on the nominee purchaser within *21 days* of the date of service of the initial notice, requiring the nominee purchaser to deduce the title of any qualifying tenant supporting the notice.

- The nominee purchaser must comply with that notice within *21 days*, failing which there is a deemed withdrawal of the initial notice.

- The freeholder (and any other relevant landlord) has the right of access to the building on not less than *10 days'* written notice.

- The freeholder must serve a counternotice within *two months* of the initial notice (or such longer period as is allowed in the initial notice).

- If no counternotice (or no valid counternotice) is served within the time period, the nominee purchaser *must* apply to the county court within *six months* for an order to purchase the freehold, otherwise the nominee purchaser is deemed to have withdrawn the initial notice and the tenants are prohibited from serving a fresh initial notice for a period of 12 months.

- If the freeholder disputes the tenants' rights, the tenants have *two months* from the date of the counternotice to apply to court for a declaration as to their right to enfranchise.

- If the freeholder admits the tenants' rights, then between *two and six months* from the date of the counternotice the nominee purchaser *must* apply to the Leasehold Valuation Tribunal for a determination of the terms of the acquisition (unless previously agreed between the parties), otherwise the usual deemed withdrawal and 12-month embargo on serving a fresh initial notice will apply.

## Ascertaining the reversioner
5-25

The landlord's side of the claim is conducted by the "reversioner" who is usually the freeholder of the premises. The reversioner acts on behalf of all the intermediate interests. Schedule 1 of the 1993 Act sets out the court's powers to appoint an alternative reversioner; and sections 26 and 27 provide for an application to be made to court where the freeholder cannot be found (note that this requires the support of two-thirds of the qualifying tenants) and the court may (if it is satisfied that the tenants' claim is valid) make a vesting order by referring the matter back to the Leasehold Valuation Tribunal as to the terms of the vesting.

Section 11 gives any qualifying tenant the right to obtain information about superior interests. The notice requesting this information may be served on the immediate landlord or on anyone receiving the rent and the recipient must respond within 28 days.

## Initial notice
5-26

The initial notice of the tenants' claim is the first step in the exercise of their rights and must be given by the tenants of *not less than half* of the

## Collective Enfranchisement Rights

total number of flats in the building. All of the tenants giving the notice must be qualifying tenants. The date of the initial notice becomes the valuation date, and also triggers the tenants' liability for the landlord's costs. There is no prescribed form of notice but it must comply with the requirements of section 13 of the Act and, in particular, it must:

- set out details of the freehold property and any additional premises sought

- give details of any leasehold interests to be acquired, such as an intervening head lease, and any units subject to mandatory leaseback to the freeholder

- state what additional rights or interests over any property the tenants seek

- attach a plan showing both the premises to be acquired and the property over which rights are sought

- contain a statement that the premises qualify for the right of collective enfranchisement on the relevant date, and how they qualify

- set out the full names and addresses of all the qualifying tenants in the premises and sufficient details of their leases to show that they are long leaseholders

- offer a realistic price for the freehold (including the price for any intermediate interests between the freehold and the qualifying tenants)

- specify the identity of the nominee purchaser (who will hold the freehold on the tenants' behalf, each tenant retaining that tenant's individual flat lease)

- give a date for service of the landlord's counternotice (at least two months hence)

- be signed by *all* the supporting qualifying tenants (it is not enough for the nominee purchaser to sign on their behalf).

## 5-27
Failure to specify a realistic price for the freehold may lead to the initial notice being held invalid: see *Cadogan* v *Morris* [1999] 1 EGLR 59, in which a proposal of £100 made the notice invalid where a realistic figure would have been between £100,000 and £300,000. If the landlord fails to serve a valid counternotice in time, then the tenants may be entitled to the property on the terms proposed by the tenants in their initial notice (hence the importance of the tenants specifying a realistic price): see *Willingale* v *Globalgrange Ltd* [2000] 2 EGLR 55.

## 5-28
If the notice is defective, *Sinclair Gardens Investments (Kensington) Ltd* v *Poets Chase Freehold Co Ltd* [2007] EWHC 1776 (Ch); [2007] 32 EG 89 (CS) makes it clear that the notice is considered a nullity and so the 12-month bar on serving a fresh notice does not apply.

## 5-29
The notice must be served on the freeholder and on any other relevant landlord and may be served by post (section 99).

## Effect of notice
### 5-30
Service of the initial notice triggers the procedural timetable (see Flowchart F, para 5-60) and the qualifying tenants' liability for the landlord's costs. It is also the valuation date.

## 5-31
Any claim already made by an individual tenant for a lease extension in relation to that tenant's own flat (permitted under the 1993 Act) is suspended pending the outcome of the enfranchisement claim; similarly, any binding contract for the disposal of a reversionary interest in the premises may also be suspended (unless the contract specifically provides for service of such a notice).

## 5-32
Where a freehold disposal is proposed under the 1987 Act after an initial notice has been served under the 1993 Act, there is no express provision as to what should happen, but in practice, the tenants will

normally proceed to acquire the freehold under the 1987 Act rather than under the 1993 Act, leaving their initial notice in abeyance. This is because a purchase under the 1987 Act has the following advantages:

- The purchase price is specified in the landlord's notice, so the tenants do not need to obtain expert valuation advice in order to ensure that their notice specifies a realistic price.

- Because the purchase price is certain, the tenants can immediately know whether or not they can afford it.

- The tenants do not need to pay the landlord's costs in the case of an acquisition under the 1987 Act (unless they withdraw in certain circumstances).

- There is a specific and relatively tight timetable leading to exchange of contracts under the 1987 Act.

- A larger group of tenants will be qualifying tenants for the purposes of the 1987 Act (but not for the purposes of the 1993 Act), such as Rent Act tenants.

### 5-33
The tenants should protect the initial notice by registering a unilateral notice (or an agreed notice, if the landlord agrees to the registration) at HM Land Registry, or by registering a land charge at the Land Charges Registry where the land is unregistered. Registration places various restrictions on the disposal of various reversionary interests, and binds any purchaser of the freehold.

## Counter-notice
### 5-34
The freeholder must, within the time stated in the tenants' initial notice (which must be not less than two months from the date of service of the tenants' initial notice), serve a counternotice in accordance with the 1993 Act.

### 5-35
Again, there is no prescribed form of notice, but in the notice the freeholder must do the following:

- State whether or not the claim is agreed.

- If the claim is not agreed:
  - give reasons for disputing the validity of the claim or the validity of the notice, or
  - claim an intention to redevelop the whole or a substantial part of the premises (this opposition to the claim is only available if at least two-thirds of the leases are due to terminate within five years of the date of the leaseholders' initial notice, and the landlord cannot reasonably redevelop without obtaining possession of the building).

- Give an address for service in England and Wales.

- State whether the premises are within the area of a scheme approved as an estate management scheme (see Chapter 6, para 6-16–6-21).

## 5-36
In addition, if the freeholder admits the claim, the freeholder must:

- set out which of the proposals are agreed, and which are not, and set out any counterproposals, eg as to price, extent of property to be conveyed, etc

- state whether a leaseback of any non-qualifying units such as commercial units or assured shorthold tenancies is required, as failure to seek a leaseback at this stage is fatal for the landlord (see para 5-44 below)

- give details of any other property that the freeholder requires the nominee purchaser to take

- set out details of any lease extension notice received from an individual tenant and any counternotice served in response

- give details of any other property owned by the freeholder or by another landlord which would be of no practical benefit to either of them or would cease to be capable of being managed by them and which they require the nominee purchaser to take.

## 5-37
Any counterproposal as to price need not be realistic in order for the counternotice to be valid: *9 Cornwall Crescent London Ltd* v *Kensington and Chelsea Royal London Borough Council* [2005] EWCA Civ 324; [2005] 14 EG 128 (CS).

## 5-38
In *7 Strathray Gardens Ltd* v *Pointstar Shipping & Finance Ltd* [2004] EWCA Civ 1669; [2005] 01 EG 95 (CS), the landlord failed to state in its counternotice whether the premises were in the area of an estate management scheme. The premises were not in fact in such an area, and so the counternotice should have contained a negative statement to that effect. However, the failure to make this statement was not fatal as the requirement to make it was not mandatory. There was no possible prejudice to the lessees or the nominee purchaser as a result of the absence of the information.

## 5-39
The need specifically to state whether or not the tenants' rights are admitted was highlighted in the individual lease extension case of *Burman* v *Mount Cook Land Ltd* [2001] EWCA Civ 1712; [2001] 48 EG 128 (CS). However, a failure by the landlord to specify his own name correctly was not fatal in another individual lease extension case, that of *Lay* v *Ackerman* [2004] EWCA Civ 184; [2005] 1 EGLR 139.

## 5-40
Failure to serve a counternotice entitles the tenants to apply to the county court for an order to acquire the freehold on the terms they specified: see also *Willingale* v *Globalrange Ltd*.

## Leaseback provisions
## 5-41
The tenants' nominee purchaser is required to grant one or more leases back to the former freeholder in certain circumstances under Schedule 9, Part III of the 1993 Act, such as some flats let under secure tenancies and flats let under non-qualifying tenancies by housing associations.

## 5-42

In addition, the former freeholder may require, in the counternotice, the leaseback of any unit not let to a qualifying tenant, such as a commercial unit or a flat let on an assured shorthold tenancy. The grant of such a lease takes place immediately after the transfer of the freehold. The detailed terms of the lease are laid down in Schedule 9, Part IV, but in summary the lease will be a 999-year lease at a peppercorn rent containing the following provisions:

- Repair of the structure and exterior by the landlord (the new freeholder).
- Insurance by the landlord.
- Internal repairing obligations by the tenant (the former freeholder).
- Contribution to costs such as repair and insurance to be made by the tenant.
- Landlord's consent required to assign, sublet in the case of a business lease, but otherwise the lease to be freely alienable.
- No break clause within the lease.

## 5-43

Any deviation from such provisions must be by agreement and with the consent of the Leasehold Valuation Tribunal.

## 5-44

It had been thought that it was possible to request a leaseback at any time before the transfer of the freehold, eg after valuation. This possibility has now been rendered impossible following the Court of Appeal decision in *Cawthorne v Hamdan* [2007] EWCA Civ 6; [2007] 05 EG 306 (CS). The landlord failed to seek a leaseback of the one non-qualifying flat in a block of six flats in his counternotice, and indeed completed the counternotice saying that there were no leaseback proposals. The parties were unable to agree a price for the freehold, and this was determined by the Leasehold Valuation Tribunal. A substantial part of the price consisted of the value of the non-qualifying flat which the freeholder would obviously be able to sell on a long lease at a substantial premium. The tenants then discovered that there was an agreement between the occupier of the flat and the landlord which would affect the price, and appealed to the Lands Tribunal. Almost on the eve of the hearing, the landlord required a leaseback of the flat. The

Court held that this was too late; the landlord should have sought the leaseback in his counternotice.

## Next steps
### 5-45
Any claim as to the validity of the notice or the tenants' rights, or as to the landlord's opposition on the grounds of redevelopment is dealt with by the county court.

### 5-46
If the freeholder disputes the tenants' right to acquire the freehold, the nominee purchaser must apply to the county court within two months of the counternotice for a declaration that the notice is valid and the tenants are entitled to the freehold. If no application is made in time, the initial notice is deemed to have been withdrawn, and the tenants are then barred from serving a fresh notice for 12 months.

### 5-47
Where the freeholder opposes the initial notice on the ground of redevelopment, the freeholder has two months from the date of service of the counternotice to apply to the county court for an order that the right to enfranchise shall not be exercisable, otherwise the freeholder must then serve an unopposed counternotice, failing which the nominee purchaser may apply to the county court for an order that it purchase the freehold on the terms set out in its initial notice.

### 5-48
Any dispute as to the terms of the acquisition, or as to compliance with the conveyancing timetable, is dealt with by the Leasehold Valuation Tribunal.

### 5-49
If the freeholder admits the right to enfranchise but not the terms proposed by the tenants, the parties then negotiate or, within two—six months from the date of the freeholder's counternotice, apply to the Leasehold Valuation Tribunal for determination of the terms of the acquisition. Where the nominee purchaser fails to make that application,

the usual deemed withdrawal and embargo on serving a fresh initial notice for 12 months apply.

**5-50**
The parties are expected to enter into the contract within two months of either the agreement of the terms of acquisition or their determination. Failure to do so means that either party may apply within this period to the Leasehold Valuation Tribunal for an order that the freehold be vested in the nominee purchaser on the terms agreed between the parties (or determined by the tribunal). Failure by the nominee purchaser to make the application within time leads to the usual deemed withdrawal (and its consequences).

## *Conveyance*

**5-51**
The timetable for delivery of the freeholder's title, and for the drafting and agreement of the contract is set out in Schedule 1 to the Leasehold Reform (Collective Enfranchisement and Lease Renewal) Regulations 1993 (SI 1993/2407). The contents of the conveyance are set out in Schedule 7 to the 1993 Act.

**5-52**
In brief, the nominee purchaser can require the freeholder to deduce title at any time after the counternotice is served (or time for service has expired) and the freeholder must comply within 28 days. There is then a period of 14 days for the raising of requisitions and 14 days for replies.

**5-53**
Once the terms of the acquisition have been ascertained, the freeholder must submit a draft contract (with draft transfer attached) within 21 days of agreement or determination, and there is a period of 14 days for approval/amendment (failing which there is deemed acceptance of the documents as drawn) and a further 14-day period for re-amendment/ approval (failing which, there is again a deemed acceptance, this time of the amendments made).

## 5-54
Care should be taken in moving straight to conveyance as the time limits detailed above are dependent on the process being exchange of contracts followed by conveyance.

## 5-55
The conveyance will deal with the rights to be included or reserved, including not only rights already reserved to the freeholder under the terms of the leases for the benefit of other properties, but also any additional rights which may be necessary for the reasonable enjoyment of other premises in which the freeholder has an interest.

# Valuation

## 5-56
The price paid to the freeholder on enfranchisement is ascertained (Schedule 6, Part II of the 1993 Act) by looking at the aggregate of:

- the value of the freeholder's interest on the open market (subject to certain assumptions)

- the freeholder's share of the marriage value (the amount by which the value of the unencumbered freehold exceeds the aggregate of the value of the freehold with leases and the value of the individual leases)

- compensation payable to the freeholder (for other losses such as diminution in value to another property by compulsory sale of this property by, for example, loss of development potential).

## 5-57
The freeholder's share of marriage value is now fixed at 50%. Marriage value is defined in Schedule 6 as including the power of the participating leaseholders to grant to themselves new 999-year leases at a peppercorn rent and at a nil premium. However, where at the date of the initial notice the unexpired term of any lease held by a participating tenant exceeds 80 years, any marriage value attributable to that lease is to be ignored under Schedule 6, para 4(1).

## 5-58
Schedule 6, Part III of the Act contains provisions for calculating the price payable for intermediate leasehold interests.

## 5-59
Valuation under the 1993 Act is extremely complex and advice needs to be taken from an expert valuer. The position has recently become more complex—and the cost of collective enfranchisement has in many cases become higher for tenants—as a result of the House of Lords decision in *Cadogan* v *Sportelli* [2008] UKHL 71; [2008] 50 EG 72 (CS).

*Summary of Collective Enfranchisement Rights*

# Flowchart F: Collective enfranchisement; timetable and consequences

**5-60**

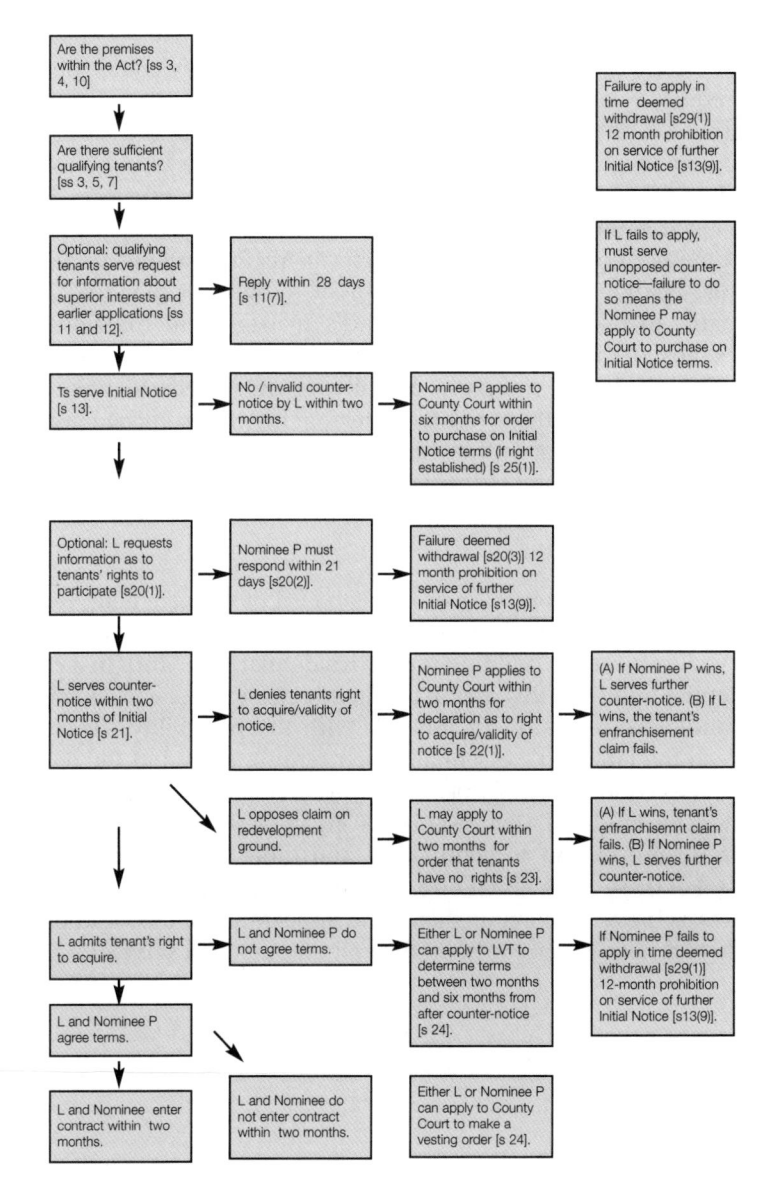

```
┌─────────────────────┐
│ Are the premises    │
│ within the Act? [ss │
│ 3, 4, 10]           │
└──────────┬──────────┘
           ▼
┌─────────────────────┐                                    ┌─────────────────────┐
│ Are there sufficient│                                    │ Failure to apply in │
│ qualifying tenants? │                                    │ time deemed         │
│ [ss 3, 5, 7]        │                                    │ withdrawal [s29(1)] │
└──────────┬──────────┘                                    │ 12 month prohibition│
           ▼                                               │ on service of further│
┌─────────────────────┐                                    │ Initial Notice      │
│ Optional: qualifying│                                    │ [s13(9)].           │
│ tenants serve req.  │   ┌──────────────────┐             └─────────────────────┘
│ for information abt │──▶│Reply within 28   │
│ superior interests  │   │days [s 11(7)].   │             ┌─────────────────────┐
│ and earlier applics │   └──────────────────┘             │ If L fails to apply,│
│ [ss 11 and 12].     │                                    │ must serve          │
└──────────┬──────────┘                                    │ unopposed counter-  │
           ▼                                               │ notice—failure to do│
┌─────────────────────┐   ┌──────────────────┐  ┌────────────────┐  so means the
│ Ts serve Initial    │   │ No / invalid     │  │Nominee P applies│  Nominee P may
│ Notice [s 13].      │──▶│ counter-notice by│─▶│to County Court  │  apply to County
└──────────┬──────────┘   │ L within two     │  │within six months│  Court to purchase
           │              │ months.          │  │for order to     │  on Initial Notice
           │              └──────────────────┘  │purchase on      │  terms.
           │                                    │Initial Notice   │
           │                                    │terms (if right  │
           │                                    │established)     │
           │                                    │[s 25(1)].       │
           │                                    └─────────────────┘
           ▼
┌─────────────────────┐   ┌──────────────────┐  ┌────────────────┐
│ Optional: L requests│   │Nominee P must    │  │Failure deemed  │
│ information as to   │──▶│respond within 21 │─▶│withdrawal      │
│ tenants' rights to  │   │days [s20(2)].    │  │[s20(3)] 12     │
│ participate [s20(1)]│   └──────────────────┘  │month prohibit. │
└──────────┬──────────┘                         │on service of   │
           ▼                                    │further Initial │
┌─────────────────────┐   ┌──────────────────┐  │Notice [s13(9)].│
│ L serves counter-   │   │L denies tenants  │  └────────────────┘
│ notice within two   │──▶│right to acquire/ │  ┌────────────────┐ ┌────────────────┐
│ months of Initial   │   │validity of notice│  │Nominee P applies│ │(A) If Nominee P│
│ Notice [s 21].      │   └──────────────────┘─▶│to County Court  │─▶│wins, L serves  │
└──────────┬──────────┘                         │within two months│ │further counter-│
           │                                    │for declaration  │ │notice. (B) If L│
           │                                    │as to right to   │ │wins, tenant's  │
           │              ┌──────────────────┐  │acquire/validity │ │enfranchisement │
           │              │L opposes claim on│  │of notice        │ │claim fails.    │
           │          ───▶│redevelopment     │  │[s 22(1)].       │ └────────────────┘
           │              │ground.           │  └─────────────────┘
           │              └──────────────────┘  ┌────────────────┐ ┌────────────────┐
           │                                    │L may apply to  │ │(A) If L wins,  │
           │                                    │County Court    │ │tenant's        │
           │                                    │within two      │─▶│enfranchisemnt  │
           │                                    │months for order│ │claim fails. (B)│
           │                                    │that tenants    │ │If Nominee P    │
           │                                    │have no rights  │ │wins, L serves  │
           │                                    │[s 23].         │ │further counter-│
           ▼                                    └────────────────┘ │notice.         │
┌─────────────────────┐   ┌──────────────────┐                     └────────────────┘
│ L admits tenant's   │   │L and Nominee P   │  ┌────────────────┐ ┌────────────────┐
│ right to acquire.   │   │do not agree terms│  │Either L or     │ │If Nominee P    │
└──────────┬──────────┘   └──────────────────┘─▶│Nominee P can   │─▶│fails to apply  │
           ▼                                    │apply to LVT to │ │in time deemed  │
┌─────────────────────┐                         │determine terms │ │withdrawal      │
│ L and Nominee P     │                         │between two     │ │[s29(1)] 12-    │
│ agree terms.        │                         │months and six  │ │month prohibit. │
└──────────┬──────────┘                         │months from     │ │on service of   │
           ▼                                    │after counter-  │ │further Initial │
┌─────────────────────┐   ┌──────────────────┐  │notice [s 24].  │ │Notice [s13(9)].│
│ L and Nominee enter │   │L and Nominee do  │  └────────────────┘ └────────────────┘
│ contract within two │   │not enter contract│  ┌────────────────┐
│ months.             │   │within two months.│─▶│Either L or     │
└─────────────────────┘   └──────────────────┘  │Nominee P can   │
                                                │apply to County │
                                                │Court to make a │
                                                │vesting order   │
                                                │[s 24].         │
                                                └────────────────┘
```

209

## B. The Commonhold and Leasehold Reform Act 2002—the right to manage

### 5-61
Since the Landlord and Tenant Act 1987, leaseholders have had the right to apply to the court for the appointment of a manager, and even compulsorily to acquire the landlord's interest, if they could establish the required landlord default. The Commonhold and Leasehold Reform Act 2002, by contrast, allows tenants to take over the management of their block of flats (or, in the case of a mixed use building, the residential part of that building) without any necessity for showing default on the landlord's or management company's part. The Commonhold and Leasehold Reform Act 2002 allows qualifying tenants, acting together in the form of a right to manage (RTM) company, to take over their landlord's management functions and places most of the landlord's duties and liabilities under the residential leases on the RTM company. Initially, tenants seem to have been reluctant to exercise these rights, preferring instead to enfranchise, but the increased costs of enfranchisement following *Cadogan* v *Sportelli* may well lead to greater enthusiasm for exercising the right to manage.

### 5-62
The landlord loses control over the standard of management, repairs, major works, and improvements, provision of services, budgets and reserve funds for the building (or the residential part of it). In a purely residential building, this may have an adverse impact on the value of the landlord's reversionary interest, with the possibility for conflict between the landlord's essentially long term interest and the tenants' sometimes shorter term focus. However, in a mixed use building, the effect on the landlord's interest may be greater even though the tenants' rights do not extend to commercial and other non-residential parts (control of these areas remaining with the landlord) with regard to such matters as the overall maintenance and appearance of the building as a whole and maintenance of shared common parts and services.

## Legislation

**5-63**
The tenants' rights are set out in:

- Sections 71–113 of the Commonhold and Leasehold Reform Act 2002 are the primary source of the tenants' rights.

- Schedule 6 sets out details of the premises excluded from the rights.

- Schedule 7 deals with ancillary matters.

- The Right to Manage (Prescribed Particulars and Forms) (England) Regulations 2003 (SI 2003/1988) gives details of the prescribed forms of key notices and the information required in them.

## Qualification

**5-64**
The qualification provisions for buildings and qualifying tenants are the same as for collective enfranchisement (except that there is no exclusion of tenants who have three or more flat leases)—see paras 5-06– 5-20 above.

## Procedure

**5-65**
The steps to be taken by the tenants and the timetable to be followed are set out in Flowchart G, see para 5-84 below.

### The RTM company
**5-66**
The first step in the exercise of their rights by the tenants is the formation of the RTM company. Sections 73 and 74 of the 2002 Act set out the requirements for the company, its membership and regulations. It must be a private company, limited by guarantee and its memorandum and articles of association must comply with the RTM Companies (Memorandum and Articles of Association) (England) Regulations 2003 (SI 2003/2120).

## Notice of invitation to participate
**5-67**
All qualifying tenants are entitled to join the RTM company, as is the freeholder and any intermediate leaseholder(s) between the freeholder and the flat leaseholders, and accordingly the RTM company must formally invite every qualifying tenant in the building to become a member of the company. Notice must be given in a prescribed form (see the Right to Manage (Prescribed Particulars and Forms) (England) Regulations 2003 (SI 2003/1988)) and must be served at least 14 days before the claim notice can be served on the landlord.

**5-68**
The invitation to participate must:

- state that the RTM company intends to take over the landlord's management functions in relation to the building (or the residential part of it)

- give the names and addresses of the members of the RTM company

- state whether the RTM company intends to appoint a managing agent, and if so who, and if not give details of the management experience of the relevant members of the RTM company

- attach a copy of the memorandum and articles of association of the RTM company.

## Notice of claim
**5-69**
The next step for the tenants is service by the RTM company of its notice of claim to manage. At this stage, the membership of the RTM company must comprise not less than 50% of the total number of tenants in the building, and all the members must be qualifying tenants.

**5-70**
The notice must be served in a prescribed form on the landlord, giving the landlord a minimum of one month to respond. The notice also gives a further date, being not less than three months after the first date, which is the date on which the RTM company intends to take

over management. It is possible, if unopposed, that the RTM company could take over management from the landlord only four months after service of this notice.

## Effect of notice of claim
**5-71**
Service of a notice of claim entitles the RTM company to request access to the building to inspect plant, the fabric of building, etc on giving at least 10 days' notice to the landlord.

**5-72**
Service of a notice of claim also obliges the RTM company to pay the landlord's reasonable costs in connection with the right to manage claim, excluding the costs of any hearing before the Leasehold Valuation Tribunal where the landlord loses.

**5-73**
Finally, service of a notice of claim obliges the landlord to serve a counternotice, otherwise the landlord is deemed to have accepted the claim.

**5-74**
Any disputes as to the RTM company's notice of claim or as to the counternotice will be determined by the Leasehold Valuation Tribunal.

## Landlord's counter-notice
**5-75**
This must also be in a prescribed form and served by the date specified in the notice of claim. The landlord can only oppose on one or more of the following grounds:

- The building does not qualify.
- The RTM company does not comply with legislation.
- The RTM company has insufficient members.

**5-76**
The landlord cannot raise queries or dispute the notice on any other ground.

## Request for information
### 5-77
In addition, the landlord may receive from the tenants or the RTM company (before or after service of the notice of claim) requests for information:

- Under section 25 of the Landlord and Tenant Act 1985, details of the landlord may be sought. The landlord has 21 days to comply, and failure to do so is a criminal offence attracting a maximum fine of £2,500.

- Under section 82 of the Commonhold and Leasehold Reform Act 2002, details of intermediate/superior interests, current arrears, insurance and management arrangements, details of all contracts in force concerning the building, other leaseholders' details, and information as to the overall state of repair (including identification of potential major works, surveys and technical reports) can all be sought by the tenants.

### 5-78
The landlord must respond within 28 days. These requests may be the first indication the landlord has of the proposed right to manage application.

## Next steps
### 5-79
The transfer of the landlord's right to manage to the RTM company occurs on the acquisition date. This is:

- the date of acquisition given in the notice of claim, where the claim is undisputed
- three months from the landlord's later agreement to the claim
- three months from the Leasehold Valuation Tribunal's determination.

### 5-80
Before the RTM company acquires the management of the building, the landlord must serve on each contractor a "contractor notice" under section 92 of the 2002 Act, stating that the right to manage is to be transferred and advising the contractor to contact the RTM company.

These notices should be served as soon as reasonably practicable after the determination date, which is three months before the acquisition date. The landlord must also serve a "contract notice" on the RTM company, setting out the details of the contracts that the landlord has for maintenance of the building and its services, the timescale being the same as for the contractor notices. The RTM company may also serve a further information request on the landlord, seeking details of contracts, building accounts, maintenance schedules, etc and the landlord must respond within 28 days (but not before the acquisition date).

### 5-81
On the acquisition date, or as soon as reasonably practicable thereafter, the landlord must transfer to the RTM company advance payments on account of service charges not yet spent and reserve/sinking funds. An external audit would be useful to prevent disputes. Rents remain the landlord's at all times.

## Landlord's membership of RTM company

### 5-82
The landlord has a right to be a member of the RTM company. Ascertaining the number of votes the landlord has can be somewhat complicated, but basically:

- The landlord has one vote for each retained flat.

- If the landlord has no retained flat(s), the landlord has one vote.

- Intermediate landlords and split freeholders are also entitled to be members of the RTM company. In such cases, the votes are allocated pro rata—so if there are two landlord members (with no retained flats) then each landlord will have one vote, and accordingly each tenant will have two votes per flat.

- If there is only one landlord but the landlord has four flats, then the landlord has four votes but each tenant has only one vote per flat.

- The landlord also has a vote for commercial units (this is because even though the RTM company does not manage the commercial units, the overall management of the building will impact on the commercial units). There is a complex multiplier to ascertain how many votes the landlord receives for the commercial units based on multiplying the number of flats by the non-residential floor area, then dividing by the total floor area.

## Problem areas

**5-83**
In spite of his membership of the RTM company, problems remain for the landlord of mixed use properties:

- *Commercial/other non-residential parts*: These remain the landlord's responsibility, but there is no provision in the 2002 Act for resolving disputes as to, for example, the external appearance of the building, correction of inherent defects that extend across both the residential and non-residential parts of the building, access, maintenance of common parts used by both residential and commercial leaseholders, signage (where this is on the residential parts), etc.

- *Forfeiture*: This remains a right of the landlord only and so the RTM company may have problems recovering arrears.

- *Non-management covenants*: These remain the landlord's responsibility. The RTM company must ensure that leaseholders comply with their covenants and notify the landlord of any default.

- *Approvals*: These are now the responsibility of the RTM company, but notice must be given to the landlord who may object, leading to a Leasehold Valuation Tribunal determination ultimately.

- *Contracts*: There is no express provision in the 2002 Act that brings to an end any fixed term management and maintenance contracts originally entered into between the landlord and parties such as lift maintenance companies, entryphone system operators and cleaning service companies. If these parties decline to deal with the RTM company (or the RTM company decides to enter into fresh contracts with third parties), the landlord may potentially be liable for breach of the contracts and may be required to continue making payments under those contracts without being able to recover those payments from the tenants. It appears that the original intention of Parliament may have been for these contracts to be frustrated (and therefore come to an end) on the RTM company acquiring the landlord's management functions, but it is not clear that this would in fact be the legal position. It has also been suggested that the landlord may have a claim against the tenants or the RTM company for being deprived of the right to recoup the cost of compliance with the contracts, but this is not provided for in the legislation.

# Flowchart G: Right to manage timetable
## 5-84

- Are the premises within the Act? [s72 and Sch 6].
- Are there sufficient qualifying tenants? [ss 72, 75, 76 and 77].
- Set up right to manage (RTM) company.
- Serve Notice of invitation to participate on all qualifying tenants [s78].
- At least 14 days later serve Notice of Claim on L [s79].
- Optional: RTM companies right to obtain information [s82] and RTM companies'/L's right to access [s83].
  - Information within 28 days [s82(3)] and access within 10 days [s83(3)].
- L serves counter notice within time specified in Notice of Claim (at least one month) [s84].
  - L admits claim (Failure to serve counter notice = deemed admission of claim [s90(3)]).
    - Acquisition date = date of Notice of Claim (at least three months after date for service of counter-notice) [s90].
      - L must serve contractor notices and contract notices as soon as reasonably practicable after acquisition date [ss 91 & 92].
- L not admit claim.
- RTM company may apply to LVT for determination as to entitlement within 2 months of counter notice [s84(3), (4)].
  - Failure to apply in time – deemed withdrawal [s87(1)] – joint and several costs liability of RTM company and members [ss88, 89].
- (A) If RTM co wins – acquisition date = 3 months from final determination [s90(4)] (B) If L wins RTM co's claim fails.
  - L must serve contractor notices and contract notices as soon as reasonably practicable after acquisition date [ss 91 & 92].

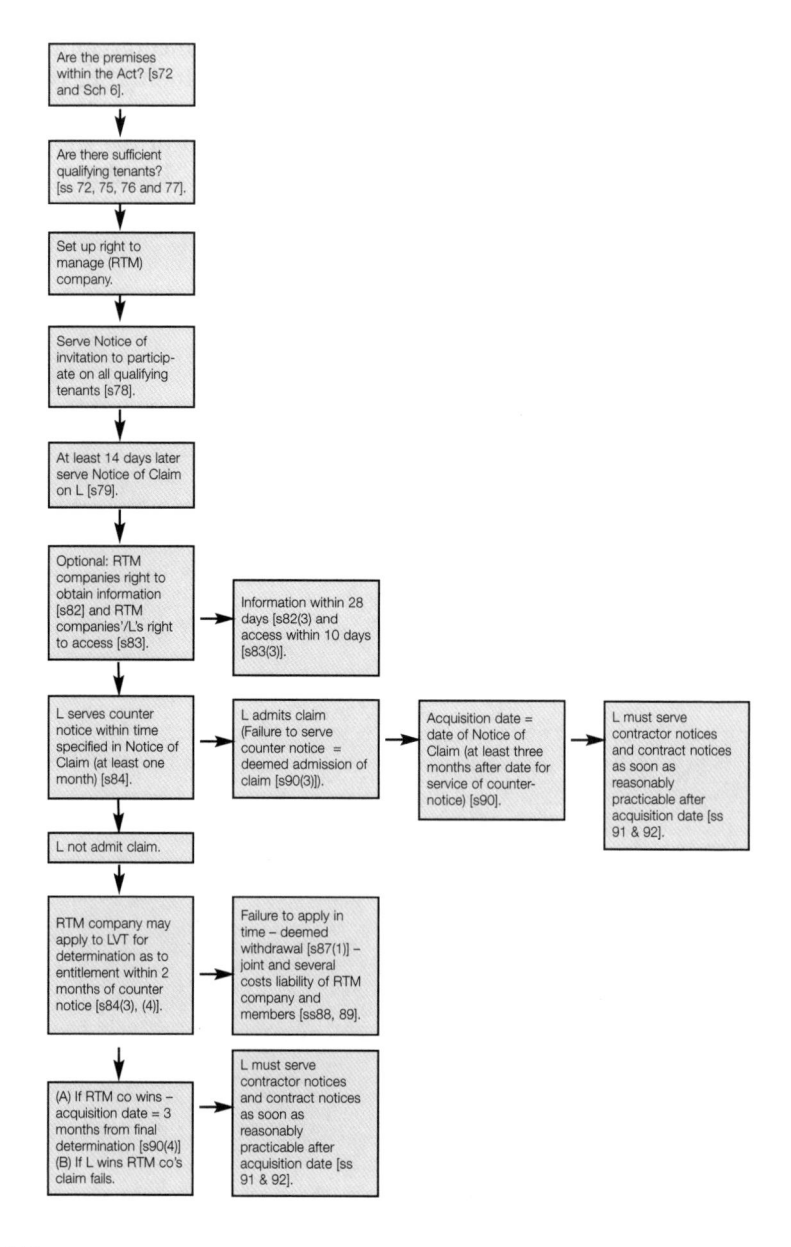

# Structuring to Avoid the Need for Compliance with the Acts

## A. The Landlord and Tenant Act 1987

**6-01**
As will be readily appreciated, having to comply with the Landlord and Tenant Act 1987 can be time consuming and costly for a landlord, and can therefore have an adverse effect on the value of the landlord's property. The landlord may not be free to sell a property within the desired timescale: at least two months must elapse after serving section 5 notices before the landlord can proceed to exchange contracts with a third party, even if the tenants fail to respond to the section 5 notices. In the case of a proposed sale at auction, at least four months' notice must be given to the tenants, which effectively cancels out the benefits of selling property at auction.

**6-02**
A delay of two months or more may be particularly problematic where property values are falling rapidly; by the time the landlord is free to sell at the proposed price, the open market value will have reduced (so that the price that a third party purchaser will be prepared to pay will have reduced commensurately) and further time must elapse whilst a second set of section 5 notices are served at the reduced price. The same problems will arise in the case of a sale by a mortgagee in possession, which must adversely affect the landlord's ability to mortgage the property.

**6-03**
For this reason, the landlord will wish to avoid having to comply with the 1987 Act wherever possible. There are a number of methods that can be used, as follows:

## Take the premises outside the Act

**6-04**
The landlord is not required to comply with the 1987 Act where:

- The premises of which the landlord proposes to dispose do not form part of a "building". As we have seen, the Act does not contain any definition of "building", but the normal meaning of a building is a structure surrounded by walls with a roof on top. Where, therefore, bare land is to be disposed of, compliance with the Act will not be required.

  Although there is no case law on the point, it is possible that the disposal of a development site *before* the structures on that site have become "buildings" may be outwith the Act.

- The building of which the premises to be disposed of form part does not contain two or more flats. Where the building contains only a single flat, the Act will not apply to the disposal. Further, the definition of "flat" requires that the flat premises be severed horizontally from another part of the building: a single house converted into two duplexes (and not containing any other separate premises) will not, therefore, contain "flats" as defined.

  Again, although there is no case law on the point, it is possible that the disposal of a development site before the building thereon contains two or more "flats" may be outwith the Act. The definition of "flat" requires that the premises be constructed or adapted for use for the purposes of a dwelling; where no kitchen or bathroom facilities have yet been installed, arguably the building does not contain any "flats".

- Less than 50% of the internal floor area (ignoring common parts) of the building of which the premises to be disposed of form part is occupied or intended to be occupied for residential purposes. So a mixed use building that contains just 50.01% commercial will be outwith the Act, whereas a mixed use building that contains 50% commercial and 50% residential will be within the Act.

## Take the landlord outside the Act

### 6-05
Where there is a lease for more than seven years (without any landlord's option to break in the first seven years of the term) intervening between the landlord and the qualifying tenants, the landlord is not obliged to comply with the Act.

### 6-06
Note, however, that the creation of such a lease in relation to a building falling within the Act will be a disposal that is caught by the Act; the lease needs to be created before the building falls within the Act.

### 6-07
Alternatively, an individual intervening lease can be created in respect of each of a number of flats within the building (thus taking advantage of the single lease exemption — see Chapter 4, para 4-47); a subsequent disposal of the whole building by the landlord need not then comply with the Act (either because the number of qualifying tenants has been sufficiently reduced, or because the landlord is not the immediate landlord of a sufficient number of the qualifying tenants). It may be feared that each such individual intervening lease could be said to be a sham and a device for avoiding the Act: but see the comments of Neuberger J (now Lord Neuberger) in the case of *National Westminster Bank Plc v Jones* [2000] EG 82 (CS) at paragraphs 37–39:

> It is equally clear, to my mind, that the mere fact that a tenancy, or any other contractual transaction, is entered into for such an artificial purpose, namely to avoid the contractual or statutory rights which a third party would otherwise enjoy, does not by any means of itself render the transaction a sham. The point was well put by Sir Thomas Bingham MR in *Belvedere Court Management Limited* v *Frogmore Developments Limited* [1997] QB 858 at 876D-17. In that case, an apparently artificial transaction entered into by the landlord of a block of flats with a Company it effectively owned significantly reduced the benefit of the rights which the tenants of the flats would otherwise have had under the provisions of the Landlord and Tenant Act 1987, The Master of the Rolls said this:
>
> 'I share the Judge's view that these arrangements were not a sham. There was no element of pretence.... The parties were not doing one thing and saying another. I would ... accept the ... view that the ... leases were an

artificial device intended to circumvent a result the Act would otherwise have brought about. But the signing of such a device did not defeat the reversioners in *Jones* v *Wrotham Park Settled Estates* [1980] AC 74 [a case involving the grant of what may be said to be an artificial tenancy to improve the landlord's entitlement to compensation under the leasehold enfranchisement legislation] nor the lessor in *Hilton* v *Plustitle Limited* [1989] 1 WLR 129 [where a prospective residential occupier was required to acquire a company for the purpose of a letting so that the landlord could avoid the rent restriction legislation] and I am not for my part satisfied that in the field of real property the principles in *W T Ramsay* ... entitle the court simply to ignore or override apparently effective transactions which on their face confer an interest in land on the transferee. Many transactions between group companies may be artificial. That does not entitle the court in ordinary circumstances to treat such transactions as null.

These observations highlight a point emphasised by Mr Jourdan, namely that many artificial transactions, which are nonetheless valid, are normally rendered doubly artificial by the fact that they will involve a company itself an artificial person, whose artificiality is frequently increased, as in this case and indeed in the cases considered by the Master of the Rolls, by the fact that the company has solely been formed and acquired for the purpose of entering into the artificial transaction. Many partnerships form and own companies for tax, administrative, limited liability or other reasons, which can be said to be artificial.

Accordingly, while the palpable, and freely admitted, artificiality of the agreements in the present case cannot be doubted, it certainly does not follow that, as a result, the agreements must be shams. However, in my judgment, the fact that a particular transaction is palpably artificial is a factor which can properly be taken into account when deciding whether it is a sham. Indeed, it would seem to me to require very unusual circumstances before the court held that a transaction which was not artificial was in fact a sham. I add this. If the court were to conclude that a transaction was artificial, in circumstances where the party relying on it was contending that it was not artificial, then that might be a further reason (although certainly not a conclusive reason) for deciding that the transaction was a sham, given that a sham transaction involves a degree of dishonesty on the part of the parties involved. That is not the position here.

## Limit the number of qualifying tenants

### 6-08
Where the building of which the premises to be disposed of form part contains less than two flats let to qualifying tenants, the disposal is outwith the Act. Further, where the number of flats held by qualifying tenants is 50% (or less) of the total number of flats in the building, the disposal is outwith the Act. So the qualifying tenants must hold *more than* 50% of the flats in the building before the Act will apply.

Flats let to assured tenants (including assured shorthold tenants) will be flats let to non-qualifying tenants.

## Take the disposal outside the Act

### 6-09
The most obvious ways of taking the disposal outside the Act are for the landlord:

- To exchange contracts or grant an option or right of pre-emption before the premises fall within the Act. Completion of the sale pursuant to the contract, option or right of pre-emption is then a disposal outwith the Act.

- To find an associated company (with which the landlord has been associated for at least two years) and transfer the premises to that company (this is not a relevant disposal and so compliance with the Act is not required). The disposal of the premises to a third party can then proceed by way of a sale of the shares in the associated company (this will not be caught by the Act, as it is a disposal of shares, not a disposal of an estate or interest in land). Clearly, if the landlord does not happen to have a convenient associated company, then two years must elapse after the formation of such a company before this loophole can be used.

    In view of the comments of the Court of Appeal in the case of *Michaels* v *Harley House (Marylebone) Ltd*, the landlord would be well advised to complete the transfer to the associated company, including registration at the Land Registry (if required), before commencing the marketing of the premises to a third party by way of share sale. In that case, the landlord and the third party purchaser attempted to use the associated company and share transfer route in order to avoid having to serve offer notices on the tenants. The Court of Appeal said that the arrangement failed because the share sale started before completion of the land transfer between the associated companies, but delays on the part of the tenants (and the fact that the third party purchaser had spent considerable sums on the improvement of the property) meant that the tenants failed in their bid to acquire the building.

Is there a risk that, if the landlord and the third party act together to arrange for the disposal via an associated company in order to avoid the Act, the landlord and the third party purchaser could be liable to the qualifying tenants for damages for conspiracy to injure them by unlawful means (or for breach of statutory duty)? Following the case of *Michaels* v *Harley House (Marylebone) Ltd*, an unsuccessful claim for such a conspiracy was brought by one of the qualifying tenants in the case of *Michaels* v *Taylor Woodrow Developments Ltd* [2000] PLSCS 101.

Laddie J dismissed the action, holding that, in order to support the existence of an actionable conspiracy to injure by unlawful means, those means must be actionable in their own right against at least some of the conspirators, whereas it was conceded that the tenant had no rights against each of the conspirators individually. Originally, the tenant had sought damages also for breach of statutory duty, but this was not pursued (rightly so, in the opinion of Laddie J), as she accepted that the 1987 Act does not provide for damages for such a breach.

It should be noted, however, that the case related to the 1987 Act before it was amended by the Housing Act 1996. It is submitted that the 1996 amendments, providing for breach of the 1987 Act to be a criminal offence, makes it even less likely that an action for breach of statutory duty or for conspiracy would be successful in the future.

## 6-10

Although exchange of a conditional contract is caught by the Act, it is conceivable that the landlord's solicitor and the third party purchaser's solicitor could each hold a contract executed by the other in escrow subject to a condition that the requisite majority of qualifying tenants fails to exercise its rights under the Act. The dating of the two parts of the contract could then follow such failure, on the basis of undertakings given by each solicitor to the other.

**6-11**
It has been suggested that compliance with the Act can be avoided by the landlord granting a mortgage to a third party and then committing a breach of the obligations under the mortgage, which then entitles the third party to foreclose under the mortgage (and thus become the owner of the property). Although the creation of a mortgage (as security for a loan) is not a relevant disposal for the purposes of the Act, and a foreclosure would not appear to be a disposal of an estate or interest in land (although it is not expressly excluded as such from the definition of "relevant disposal"), foreclosure requires the leave of the court and hence is unlikely to be able to be used as a device to avoid the Act. If, however, the third party is content simply to exercise rights over the property as mortgagee in possession following the landlord's breach, without actually acquiring the property or selling it on, then control of the building will have passed to the third party without the landlord having to comply with the Act.

## Effect the disposal by way of share sale
**6-12**
If the landlord owns a building through the medium of a company (or similar body, such as a unit trust or limited liability partnership), the shares in that company (or the units or other medium through which the body is owned) can be transferred to a third party without the disposal being caught by the Act. However, the initial acquisition of the building by the company (or similar body) would need to be done at a time when the building falls outwith the Act, otherwise the disposal of the building to the company would itself be caught by the Act.

# B. The Leasehold Reform, Housing and Urban Development Act 1993 and the Commonhold and Leasehold Reform Act 2002

**6-13**
Unlike the 1987 Act, the Leasehold Reform, Housing and Urban Development Act 1993 and the Commonhold and Leasehold Reform Act 2002 do not grant rights of first refusal to residential tenants where the landlord proposes to dispose of an estate or interest in their

building, and so there is not the same need for the landlord to avoid having to comply with these Acts. The fact that a building falls within the 1993 Act and/or the 2002 Act will not affect the sale of that building by the landlord.

### 6-14

However, the landlord may not want the tenants to enjoy the rights granted to them under the 1993 Act and/or the 2002 Act. In that case, it is far more difficult for the landlord to avoid the tenants' rights arising than in the case of the 1987 Act. In particular, there are anti-avoidance provisions in the Acts, and the creation of an intervening lease will have no effect on the tenants' rights. The landlord may find it useful, however, to grant to a company controlled by the landlord, individual head-leases of the residential and non-residential parts of the mixed use building before subletting individual units; this will at least ensure that, if the freehold is purchased by way of collective enfranchisement, the landlord's head leasehold interest in the non-residential part of the building will survive, thus avoiding the need for the landlord to require a leaseback.

### 6-15

The landlord may avoid the premises falling within the 1993/2002 Acts by:

- Taking the premises outside the Acts. This will occur if the building contains fewer than two flats or is not a self-contained building. In the case of a mixed use building, if less than 75% of the building is used for residential purposes, neither Act will apply. In the case of a new development, provided that there is no "building", or the building does not yet contain at least two "flats", the legislation will not apply; but as soon as the development reaches a sufficiently advanced stage, the legislation will apply, and there are no steps that can be taken whilst the development is being carried out that can prevent the tenants from subsequently acquiring their rights, unlike in the case of the 1987 Act.

- Limiting the number of qualifying tenants. If the landlord is prepared to have a building that contains a sufficient number of non-qualifying tenants, such as assured shorthold tenants, assured tenants or Rent Act tenants, then neither Act will apply.

Note that a tenant who has acquired a flat on a "shared ownership" basis will not be a qualifying tenant until the tenant has achieved 100% ownership of the flat by "staircasing".

## C. The 1993 Act and the estate management scheme

**6-16**
Where a landlord is anxious to maintain some control over the state of repair, decorative condition and appearance of a newly-developed property, even if the leaseholders exercise their rights of collective enfranchisement under the 1993 Act, the landlord may be able to do so by way of an estate management scheme.

**6-17**
Concern about the breaking up of historic estates, and the possible deterioration of standards on such estates led to the introduction of estate management schemes under the Leasehold Reform Act 1967. Under an approved scheme a former landlord retains some control over an estate that was, or could be, subject to enfranchisement rights by reserving powers of management to the former landlord in the event of enfranchisement claims. For example, a scheme may allow a former landlord to control such matters as the frequency and manner of external painting, external alterations, development and change of use and may allow the former landlord to enter to carry out repairs. It is also possible to include within the scope of such schemes obligations on the former leaseholders to maintain and repair, and to pay any costs incurred by the former landlord in connection with enforcing the scheme.

**6-18**
The original time limits laid down in the 1967 Act for making such applications have expired. The 1993 Act provided further time limits for making applications, and those limits have also expired for most pre-existing estates. However, where the proposed application *could not* have been made before the expiry of the 1993 Act time limits, ie for estates, areas and developments not in existence before 1 November 1995, the 1993 Act provides that application for approval of a scheme may be made, on notice to all interested parties, to the Secretary of State.

## 6-19

If the Secretary of State approves such an application, the application is then dealt with by the Leasehold Valuation Tribunal, which will consider the benefit to the area as a whole, the extent to which it is reasonable to impose burdens on the former tenants and the past and present character and historical/architectural dimension of the area. The Tribunal should also ensure that the proposed scheme does not give the former landlord a disproportionate level of control, and that it provides a mechanism for ascertaining the identity of the "former landlord" in the future. The Tribunal has no power to amend a proposed scheme, but it may suggest modifications, and refuse approval for the proposed scheme should those modifications not be made.

## 6-20

Following approval, the scheme is registered as a local land charge, enforceable by the former landlord (and the former landlord's successors in title) against any person occupying or interested in any property within the scheme.

## 6-21

Once application for a scheme has been made (even where the application to the Secretary of State is pending), any enfranchisement claim is adjourned, pending the outcome of the scheme application. Any leaseholder who has made such an enfranchisement claim by that stage, may withdraw the claim (with no liability for landlord's costs) until the determination of the scheme application, and then serve a fresh enfranchisement notice immediately after the determination of the application for the proposed scheme.

# Appendix

**The Leasehold Reform, Housing and Urban Development Act 1993**
**Part I Landlord and Tenant**

## Chapter I
## Collective enfranchisement in case of tenants of flats

### Preliminary

**1 The right to collective enfranchisement**

(1) This Chapter has effect for the purpose of conferring on qualifying tenants of flats contained in premises to which this Chapter applies on the relevant date the right, exercisable subject to and in accordance with this Chapter, to have the freehold of those premises acquired on their behalf—

(a) by a person or persons appointed by them for the purpose, and
(b) at a price determined in accordance with this Chapter;

and that right is referred to in this Chapter as "the right to collective enfranchisement".

(2) Where the right to collective enfranchisement is exercised in relation to any such premises ("the relevant premises")—

(a) the qualifying tenants by whom the right is exercised shall be entitled, subject to and in accordance with this Chapter, to have acquired, in like manner, the freehold of any property which is not comprised in the relevant premises but to which this paragraph applies by virtue of subsection (3); and

(b) section 2 has effect with respect to the acquisition of leasehold interests to which paragraph (a) or (b) of subsection (1) of that section applies.

(3) Subsection (2)(a) applies to any property if . . . at the relevant date either—

(a) it is appurtenant property which is demised by the lease held by a qualifying tenant of a flat contained in the relevant premises; or
(b) it is property which any such tenant is entitled under the terms of the lease of his flat to use in common with the occupiers of other premises (whether those premises are contained in the relevant premises or not).

(4) The right of acquisition in respect of the freehold of any such property as is mentioned in subsection (3)(b) shall, however, be taken to be satisfied with respect to that property if, on the acquisition of the relevant premises in pursuance of this Chapter, either—

(a) there are granted by the person who owns the freehold of that property—

  (i) over that property, or
  (ii) over any other property,

  such permanent rights as will ensure that thereafter the occupier of the flat referred to in that provision has as nearly as may be the same rights as those enjoyed in relation to that property on the relevant date by the qualifying tenant under the terms of his lease; or

(b) there is acquired from the person who owns the freehold of that property the freehold of any other property over which any such permanent rights may be granted.

(5) A claim by qualifying tenants to exercise the right to collective enfranchisement may be made in relation to any premises to which this Chapter applies despite the fact that those premises are less extensive than the entirety of the premises in relation to which those tenants are entitled to exercise that right.

(6) Any right or obligation under this Chapter to acquire any interest in property shall not extend to underlying minerals in which that interest subsists if—

*Appendix*

(a) the owner of the interest requires the minerals to be excepted, and
(b) proper provision is made for the support of the property as it is enjoyed on the relevant date.

(7) In this section—

"appurtenant property", in relation to a flat, means any garage, outhouse, garden, yard or appurtenances belonging to, or usually enjoyed with, the flat;

. . .

"the relevant premises" means any such premises as are referred to in subsection (2).

(8) In this Chapter "the relevant date", in relation to any claim to exercise the right to collective enfranchisement, means the date on which notice of the claim is given under section 13.

## 2 Acquisition of leasehold interests

(1) Where the right to collective enfranchisement is exercised in relation to any premises to which this Chapter applies ("the relevant premises"), then, subject to and in accordance with this Chapter—

(a) there shall be acquired on behalf of the qualifying tenants by whom the right is exercised every interest to which this paragraph applies by virtue of subsection (2); and
(b) those tenants shall be entitled to have acquired on their behalf any interest to which this paragraph applies by virtue of subsection (3);

and any interest so acquired on behalf of those tenants shall be acquired in the manner mentioned in paragraphs (a) and (b) of section 1(1).

(2) Paragraph (a) of subsection (1) above applies to the interest of the tenant under any lease which is superior to the lease held by a qualifying tenant of a flat contained in the relevant premises.

(3) Paragraph (b) of subsection (1) above applies to the interest of the tenant under any lease (not falling within subsection (2) above) under which the demised premises consist of or include—

(a) any common parts of the relevant premises, or
(b) any property falling within section 1(2)(a) which is to be acquired by virtue of that provision,

where the acquisition of that interest is reasonably necessary for the proper management or maintenance of those common parts, or (as the case may be) that property, on behalf of the tenants by whom the right to collective enfranchisement is exercised.

(4) Where the demised premises under any lease falling within subsection (2) or (3) include any premises other than—

(a) a flat contained in the relevant premises which is held by a qualifying tenant,
(b) any common parts of those premises, or
(c) any such property as is mentioned in subsection (3)(b),

the obligation or (as the case may be) right under subsection (1) above to acquire the interest of the tenant under the lease shall not extend to his interest under the lease in any such other premises.

(5) Where the qualifying tenant of a flat is a public sector landlord and the flat is let under a secure tenancy [or an introductory tenancy], then if—

(a) the condition specified in subsection (6) is satisfied, and
(b) the lease of the qualifying tenant is directly derived out of a lease under which the tenant is a public sector landlord,

the interest of that public sector landlord as tenant under that lease shall not be liable to be acquired by virtue of subsection (1) to the extent that it is an interest in the flat or in any appurtenant property; and the interest of a public sector landlord as tenant under any lease out of which the qualifying tenant's lease is indirectly derived shall, to the like extent, not be liable to be so acquired (so long as the tenant under every lease intermediate between that lease and the qualifying tenant's lease is a public sector landlord).

(6) The condition referred to in subsection (5)(a) is that either—

(a) the qualifying tenant is the immediate landlord under the secure tenancy or, as the case may be, the introductory tenancy, or
(b) he is the landlord under a lease which is superior to the secure tenancy or, as the case may be, the introductory tenancy and the tenant under that lease, and the tenant under every lease (if any) intermediate between it and the secure tenancy or the introductory tenancy, is also a public sector landlord;

*Appendix*

and in subsection (5) "appurtenant property" has the same meaning as in section 1.

(7) In this section "the relevant premises" means any such premises as are referred to in subsection (1).

## 3 Premises to which this Chapter applies

(1) Subject to section 4, this Chapter applies to any premises if—

(a) they consist of a self-contained building or part of a building ...;
(b) they contain two or more flats held by qualifying tenants; and
(c) the total number of flats held by such tenants is not less than two thirds of the total number of flats contained in the premises.

(2) For the purposes of this section a building is a self-contained building if it is structurally detached, and a part of a building is a self-contained part of a building if—

(a) it constitutes a vertical division of the building and the structure of the building is such that that part could be redeveloped independently of the remainder of the building; and
(b) the relevant services provided for occupiers of that part either—
   (i) are provided independently of the relevant services provided for occupiers of the remainder of the building, or
   (ii) could be so provided without involving the carrying out of any works likely to result in a significant interruption in the provision of any such services for occupiers of the remainder of the building;

and for this purpose "relevant services" means services provided by means of pipes, cables or other fixed installations.

## 4 Premises excluded from right

(1) This Chapter does not apply to premises falling within section 3(1) if—

(a) any part or parts of the premises is or are neither—
   (i) occupied, or intended to be occupied, for residential purposes, nor
   (ii) comprised in any common parts of the premises; and

(b) the internal floor area of that part or of those parts (taken together) exceeds 25 per cent of the internal floor area of the premises (taken as a whole).

(2) Where in the case of any such premises any part of the premises (such as, for example, a garage, parking space or storage area) is used, or intended for use, in conjunction with a particular dwelling contained in the premises (and accordingly is not comprised in any common parts of the premises), it shall be taken to be occupied, or intended to be occupied, for residential purposes.

(3) For the purpose of determining the internal floor area of a building or of any part of a building, the floor or floors of the building or part shall be taken to extend (without interruption) throughout the whole of the interior of the building or part, except that the area of any common parts of the building or part shall be disregarded.

(3A) Where different persons own the freehold of different parts of premises within subsection (1) of section 3, this Chapter does not apply to the premises if any of those parts is a self-contained part of a building for the purposes of that section.

(4) This Chapter does not apply to premises falling within section 3(1) if the premises are premises with a resident landlord and do not contain more than four units.

(5) This Chapter does not apply to premises falling within section 3(1) if the freehold of the premises includes track of an operational railway; and for the purposes of this subsection—

(a) "track" includes any land or other property comprising the permanent way of a railway (whether or not it is also used for other purposes) and includes any bridge, tunnel, culvert, retaining wall or other structure used for the support of, or otherwise in connection with, track,
(b) "operational" means not disused, and
(c) "railway" has the same meaning as in any provision of Part 1 of the Railways Act 1993 (c 43) for the purposes of which that term is stated to have its wider meaning.

## 5 Qualifying tenants

(1) Subject to the following provisions of this section, a person is a qualifying tenant of a flat for the purposes of this Chapter if he is tenant of the flat under a long lease . . .

(2) Subsection (1) does not apply where—

(a) the lease is a business lease; or
(b) the immediate landlord under the lease is a charitable housing trust and the flat forms part of the housing accommodation provided by it in the pursuit of its charitable purposes; or
(c) the lease was granted by sub-demise out of a superior lease other than a long lease . . ., the grant was made in breach of the terms of the superior lease, and there has been no waiver of the breach by the superior landlord;

and in paragraph (b) "charitable housing trust" means a housing trust within the meaning of the Housing Act 1985 which is a charity within the meaning of the Charities Act 1993.

(3) No flat shall have more than one qualifying tenant at any one time.

(4) Accordingly—

(a) where a flat is for the time being let under two or more leases to which subsection (1) applies, any tenant under any of those leases which is superior to that held by any other such tenant shall not be a qualifying tenant of the flat for the purposes of this Chapter; and
(b) where a flat is for the time being let to joint tenants under a lease to which subsection (1) applies, the joint tenants shall (subject to paragraph (a) and subsection (5)) be regarded for the purposes of this Chapter as jointly constituting the qualifying tenant of the flat.

(5) Where apart from this subsection—

(a) a person would be regarded for the purposes of this Chapter as being (or as being among those constituting) the qualifying tenant of a flat contained in any particular premises consisting of the whole or part of a building, but
(b) that person would also be regarded for those purposes as being (or as being among those constituting) the qualifying tenant of each of two or more other flats contained in those premises,

then, whether that person is tenant of the flats referred to in paragraphs (a) and (b) under a single lease or otherwise, there shall be taken for those purposes to be no qualifying tenant of any of those flats.

(6) For the purposes of subsection (5) in its application to a body corporate any flat let to an associated company (whether alone or jointly with any other person or persons) shall be treated as if it were so let to that body; and for this purpose "associated company" means another body corporate which is (within the meaning of section 736 of the Companies Act 1985) that body's holding company, a subsidiary of that body or another subsidiary of that body's holding company.

**6 ...**

**7 Meaning of "long lease"**
(1) In this Chapter "long lease" means (subject to the following provisions of this section)—
(a) a lease granted for a term of years certain exceeding 21 years, whether or not it is (or may become) terminable before the end of that term by notice given by or to the tenant or by re-entry, forfeiture or otherwise;
(b) a lease for a term fixed by law under a grant with a covenant or obligation for perpetual renewal (other than a lease by sub-demise from one which is not a long lease) or a lease taking effect under section 149(6) of the Law of Property Act 1925 (leases terminable after a death or marriage [or the formation of a civil partnership]);
(c) a lease granted in pursuance of the right to buy conferred by Part V of the Housing Act 1985 or in pursuance of the right to acquire on rent to mortgage terms conferred by that Part of that Act; ...
(d) a shared ownership lease, whether granted in pursuance of that Part of that Act or otherwise, where the tenant's total share is 100 per cent; or
(e) a lease granted in pursuance of that Part of that Act as it has effect by virtue of section 17 of the Housing Act 1996 (the right to acquire).

(2) A lease terminable by notice after a death, a marriage or the formation of a civil partnership is not to be treated as a long lease for the purposes of this Chapter if—

(a) the notice is capable of being given at any time after the death or marriage of, or the formation of a civil partnership by, the tenant;
(b) the length of the notice is not more than three months; and
(c) the terms of the lease preclude both—
   (i) its assignment otherwise than by virtue of section 92 of the Housing Act 1985 (assignments by way of exchange), and
   (ii) the sub-letting of the whole of the premises comprised in it.

(3) Where the tenant of any property under a long lease . . ., on the coming to an end of that lease, becomes or has become tenant of the property or part of it under any subsequent tenancy (whether by express grant or by implication of law), then that tenancy shall be deemed for the purposes of this Chapter (including any further application of this subsection) to be a long lease irrespective of its terms.

(4) Where—

(a) a lease is or has been granted for a term of years certain not exceeding 21 years, but with a covenant or obligation for renewal without payment of a premium (but not for perpetual renewal), and
(b) the lease is or has been renewed on one or more occasions so as to bring to more than 21 years the total of the terms granted (including any interval between the end of a lease and the grant of a renewal),

this Chapter shall apply as if the term originally granted had been one exceeding 21 years.

(5) References in this Chapter to a long lease include—

(a) any period during which the lease is or was continued under Part I of the Landlord and Tenant Act 1954 or under Schedule 10 to the Local Government and Housing Act 1989;
(b) any period during which the lease was continued under the Leasehold Property (Temporary Provisions) Act 1951.

(6) Where in the case of a flat there are at any time two or more separate leases, with the same landlord and the same tenant, and—

(a) the property comprised in one of those leases consists of either the flat or a part of it (in either case with or without any appurtenant property), and

(b) the property comprised in every other lease consists of either a part of the flat (with or without any appurtenant property) or appurtenant property only,

then in relation to the property comprised in such of those leases as are long leases, this Chapter shall apply as it would if at that time—

(i) there were a single lease of that property, and
(ii) that lease were a long lease;

but this subsection has effect subject to the operation of subsections (3) to (5) in relation to any of the separate leases.

(7) In this section—

"appurtenant property" has the same meaning as in section 1;
"shared ownership lease" means a lease—

(a) granted on payment of a premium calculated by reference to a percentage of the value of the demised premises or the cost of providing them, or
(b) under which the tenant (or his personal representatives) will or may be entitled to a sum calculated by reference, directly or indirectly, to the value of those premises; and

"total share", in relation to the interest of a tenant under a shared ownership lease, means his initial share plus any additional share or shares in the demised premises which he has acquired.

8 ...

## 9 The reversioner and other relevant landlords for the purpose of this Chapter

(1) Where, in connection with any claim to exercise the right to collective enfranchisement in relation to any premises [the freehold of the whole of which is owned by the same person], it is not proposed to acquire any interests other than—

(a) the freehold of the premises, or

(b) any other interests of the person who owns the freehold of the premises,

that person shall be the reversioner in respect of the premises for the purposes of this Chapter.

(2) Where, in connection with any such claim as is mentioned in subsection (1), it is proposed to acquire interests of persons other than the person who owns the freehold of the premises to which the claim relates, then—

(a) the reversioner in respect of the premises shall for the purposes of this Chapter be the person identified as such by Part I of Schedule 1 to this Act; and
(b) the person who owns the freehold of the premises every person who owns any freehold interest which it is proposed to acquire by virtue of section 1(2)(a), and every person who owns any leasehold interest which it is proposed to acquire under or by virtue of section 2(1)(a) or (b), shall be a relevant landlord for those purposes.

(2A) In the case of any claim to exercise the right to collective enfranchisement in relation to any premises the freehold of the whole of which is not owned by the same person—

(a) the reversioner in respect of the premises shall for the purposes of this Chapter be the person identified as such by Part IA of Schedule 1 to this Act, and
(b) every person who owns a freehold interest in the premises, every person who owns any freehold interest which it is proposed to acquire by virtue of section 1(2)(a), and every person who owns any leasehold interest which it is proposed to acquire under or by virtue of section 2(1)(a) or (b), shall be a relevant landlord for those purposes.

(3) Subject to the provisions of Part II of Schedule 1, the reversioner in respect of any premises shall, in a case to which subsection (2) [or (2A)] applies, conduct on behalf of all the relevant landlords all proceedings arising out of any notice given with respect to the premises under section 13 (whether the proceedings are for resisting or giving effect to the claim in question).

(4) Schedule 2 (which makes provision with respect to certain special categories of landlords) has effect for the purposes of this Chapter.

**10 Premises with a resident landlord**

(1) For the purposes of this Chapter any premises falling within section 3(1) are premises with a resident landlord at any time if—

(a) the premises are not, and do not form part of, a purpose-built block of flats;
(b) the same person has owned the freehold of the premises since before the conversion of the premises into two or more flats or other units; and
(c) he, or an adult member of his family, has occupied a flat or other unit contained in the premises as his only or principal home throughout the period of twelve months ending with that time.

(2) . . .

(3) . . .

(4) Where the freehold of any premises is held on trust, subsection (1) applies as if—

(a) the requirement in paragraph (b) were that the same person has had an interest under the trust (whether or not also a trustee) since before the conversion of the premises, and
(b) paragraph (c) referred to him or an adult member of his family.

(4A) . . .

(5) For the purposes of this section a person is an adult member of another's family if that person is—

(a) the other's spouse or civil partner; or
(b) a son or daughter or a son-in-law or daughter-in-law of the other, or of the other's spouse or civil partner, who has attained the age of 18; or
(c) the father or mother of the other, or of the other's spouse or civil partner;

and in paragraph (b) any reference to a person's son or daughter includes a reference to any stepson or stepdaughter of that person, and "son-in-law" and "daughter-in-law" shall be construed accordingly.

(6) In this section—

...

"purpose-built block of flats" means a building which as constructed contained two or more flats.
[...]

## Preliminary inquiries by tenants

### 11 Right of qualifying tenant to obtain information about superior interests etc

(1) A qualifying tenant of a flat may give—

(a) to any immediate landlord of his, or
(b) to any person receiving rent on behalf of any immediate landlord of his,

a notice requiring the recipient to give the tenant (so far as known to the recipient) the name and address of [every person who owns a freehold interest in] the relevant premises and the name and address of every other person who has an interest to which subsection (2) applies.

(2) In relation to a qualifying tenant of a flat, this subsection applies to the following interests, namely—

(a) the freehold of any property not contained in the relevant premises—
  (i) which is demised by the lease held by the tenant, or
  (ii) which the tenant is entitled under the terms of his lease to use in common with other persons; and
(b) any leasehold interest in the relevant premises or in any such property which is superior to that of any immediate landlord of the tenant.

(3) Any qualifying tenant of a flat may give to any person who holds a freehold interest in the relevant premises a notice requiring him to give the tenant (so far as known to him) the name and address of every person, apart from the tenant, who is—

(a) a tenant of the whole of the relevant premises, or
(b) a tenant or licensee of any separate set or sets of premises contained in the relevant premises, or

(c) a tenant or licensee of the whole or any part of any common parts so contained or of any property not so contained—
  (i) which is demised by the lease held by a qualifying tenant of a flat contained in the relevant premises, or
  (ii) which any such qualifying tenant is entitled under the terms of his lease to use in common with other persons.

(4) Any such qualifying tenant may also give—

(a) to any person who owns a freehold interest in the relevant premises,
(aa) to any person who owns a freehold interest in any such property as is mentioned in subsection (3)(c),
(b) to any person falling within subsection (3)(a), (b) or (c),

a notice requiring him to give the tenant—
  (i) such information relating to his interest in the relevant premises or (as the case may be) in any such property . . ., or
  (ii) (so far as known to him) such information relating to any interest derived (whether directly or indirectly) out of that interest,

as is specified in the notice, where the information is reasonably required by the tenant in connection with the making of a claim to exercise the right to collective enfranchisement in relation to the whole or part of the relevant premises.

(5) Where a notice is given by a qualifying tenant under subsection (4), the following rights shall be exercisable by him in relation to the recipient of the notice, namely—

(a) a right, on giving reasonable notice, to be provided with a list of documents to which subsection (6) applies;
(b) a right to inspect, at any reasonable time and on giving reasonable notice, any documents to which that subsection applies; and
(c) a right, on payment of a reasonable fee, to be provided with a copy of any documents which are contained in any list provided under paragraph (a) or have been inspected under paragraph (b).

(6) This subsection applies to any document in the custody or under the control of the recipient of the notice under subsection 4—

(a) sight of which is reasonably required by the qualifying tenant in connection with the making of such a claim as is mentioned in that subsection; and
(b) which, on a proposed sale by a willing seller to a willing buyer of the recipient's interest in the relevant premises or (as the case may be) in any such property as is mentioned in subsection (3)(c), the seller would be expected to make available to the buyer (whether at or before contract or completion).

(7) Any person who—
(a) is required by a notice under any of subsections (1) to (4) to give any information to a qualifying tenant, or
(b) is required by a qualifying tenant under subsection (5) to supply any list of documents, to permit the inspection of any documents or to supply a copy of any documents,

shall comply with that requirement within the period of 28 days beginning with the date of the giving of the notice referred to in paragraph (a) or (as the case may be) with the date of the making of the requirement referred to in paragraph (b).

(8) Where—
(a) a person has received a notice under subsection (4), and
(b) within the period of six months beginning with the date of receipt of the notice, he—
  (i) disposes of any interest (whether legal or equitable) in the relevant premises [or in any such property as is mentioned in subsection (3)(c)] otherwise than by the creation of an interest by way of security for a loan, or
  (ii) acquires any such interest (otherwise than by way of security for a loan),

then (unless that disposal or acquisition has already been notified to the qualifying tenant in accordance with subsection (7)) he shall notify the qualifying tenant of that disposal or acquisition within the period of 28 days beginning with the date when it occurred.

(9) In this section—

"document" means anything in which information of any description is recorded, and in relation to a document in which information is recorded otherwise than in legible form any reference to sight of the document is to sight of the information in legible form;

"the relevant premises", in relation to any qualifying tenant of a flat, means—

(a) if the person who owns the freehold interest in the flat owns, or the persons who own the freehold interests in the flat own, the freehold of the whole of the building in which the flat is contained, that building, or
(b) if that person owns, or those persons own, the freehold of part only of that building, that part of that building;

and any reference to an interest in the relevant premises includes an interest in part of those premises.

## 12 Right of qualifying tenant to obtain information about other matters

(1) Any notice given by a qualifying tenant under section 11(4) shall, in addition to any other requirement imposed in accordance with that provision, require the recipient to give the tenant—

(a) the information specified in subsection (2) below; and
(b) (so far as known to the recipient) the information specified in subsection (3) below.

(2) The information referred to in subsection (1)(a) is—

(a) whether the recipient has received in respect of any premises containing the tenant's flat—
   (i) a notice under section 13 in the case of which the relevant claim is still current, or
   (ii) a copy of such a notice; and
(b) if so, the date on which the notice under section 13 was given and the name and address of the nominee purchaser for the time being appointed for the purposes of section 15 in relation to that claim.

(3) The information referred to in subsection (1)(b) is—

(a) whether the tenant's flat is comprised in any property in the case of which any of paragraphs (a) to (d) of section 31(2) is applicable; and
(b) if paragraph (b) or (d) of that provision is applicable, the date of the application in question.

(4) Where—

(a) within the period of six months beginning with the date of receipt of a notice given by a tenant under section 11(4), the recipient of the notice receives in respect of any premises containing the tenant's flat—
  (i) a notice under section 13, or
  (ii) a copy of such a notice, and
(b) the tenant is not one of the qualifying tenants by whom the notice under section 13 is given,

the recipient shall, within the period of 28 days beginning with the date of receipt of the notice under section 13 or (as the case may be) the copy, notify the tenant of the date on which the notice was given and of the name and address of the nominee purchaser for the time being appointed for the purposes of section 15 in relation to the relevant claim.

(5) Where

(a) the recipient of a notice given by a tenant under section 11(4) has, in accordance with subsection (1) above, informed the tenant of any such application as is referred to in subsection (3)(b) above; and
(b) within the period of six months beginning with the date of receipt of the notice, the application is either granted or refused by the Commissioners of Inland Revenue or is withdrawn by the applicant,

the recipient shall, within the period of 28 days beginning with the date of the granting, refusal or withdrawal of the application, notify the tenant that it has been granted, refused or withdrawn.

(6) In this section "the relevant claim", in relation to a notice under section 13, means the claim in respect of which that notice is given; and for the purposes of subsection (2) above any such claim is current if—

(a) that notice continues in force in accordance with section 13(11), or
(b) a binding contract entered into in pursuance of that notice remains in force, or
(c) where an order has been made under section 24(4)(a) or (b) or 25(6)(a) or (b) with respect to any such premises as are referred to in subsection (2)(a) above, any interests which by virtue of the order fall to be vested in the nominee purchaser have yet to be so vested.

## The initial notice

**13 Notice by qualifying tenants of claim to exercise right**

(1) A claim to exercise the right to collective enfranchisement with respect to any premises is made by the giving of notice of the claim under this section.

(2) A notice given under this section ("the initial notice")—

(a) must
  (i) in a case to which section 9(2) applies, be given to the reversioner in respect of those premises; and
  (ii) in a case to which section 9(2A) applies, be given to the person specified in the notice as the recipient; and
(b) must be given by a number of qualifying tenants of flats contained in the premises as at the relevant date which—
  (i) . . .
  (ii) is not less than one-half of the total number of flats so contained;

. . .

(2A) In a case to which section 9(2A) applies, the initial notice must specify—

(a) a person who owns a freehold interest in the premises, or
(b) if every person falling within paragraph (a) is a person who cannot be found or whose identity cannot be ascertained, a relevant landlord,

as the recipient of the notice.

(3) The initial notice must—

(a) specify and be accompanied by a plan showing—
  (i) the premises of which the freehold is proposed to be acquired by virtue of section 1(1),
  (ii) any property of which the freehold is proposed to be acquired by virtue of section 1(2)(a), and
  (iii) any property ... over which it is proposed that rights (specified in the notice) should be granted by him in connection with the acquisition of the freehold of the specified premises or of any such property so far as falling within section 1(3)(a);

(b) contain a statement of the grounds on which it is claimed that the specified premises are, on the relevant date, premises to which this Chapter applies;
(c) specify—
  (i) any leasehold interest proposed to be acquired under or by virtue of section 2(1)(a) or (b), and
  (ii) any flats or other units contained in the specified premises in relation to which it is considered that any of the requirements in Part II of Schedule 9 to this Act are applicable;
(d) specify the proposed purchase price for each of the following, namely—
  (i) the freehold interest in the specified premises or, if the freehold of the whole of the specified premises is not owned by the same person, each of the freehold interests in those premises,
  (ii) the freehold interest in any property specified under paragraph (a)(ii), and
  (iii) any leasehold interest specified under paragraph (c)(i);
(e) state the full names of all the qualifying tenants of flats contained in the specified premises and the addresses of their flats, and contain ... in relation to each of those tenants, ...
  (i) such particulars of his lease as are sufficient to identify it, including the date on which the lease was entered into, the term for which it was granted and the date of the commencement of the term,
  (ii) ...
  (iii) ...;
(f) state the full name or names of the person or persons appointed as the nominee purchaser for the purposes of section 15, and an address in England and Wales at which notices may be given to that person or those persons under this Chapter; and
(g) specify the date by which the reversioner must respond to the notice by giving a counter-notice under section 21.

(4) ...

(5) The date specified in the initial notice in pursuance of subsection (3)(g) must be a date falling not less than two months after the relevant date.

(6), (7) ...

(8) Where any premises have been specified in a notice under this section, no subsequent notice which specifies the whole or part of those premises may be given under this section so long as the earlier notice continues in force.

(9) Where any premises have been specified in a notice under this section and—

(a) that notice has been withdrawn, or is deemed to have been withdrawn, under or by virtue of any provision of this Chapter or under section 74(3), or
(b) in response to that notice, an order has been applied for and obtained under section 23(1),

no subsequent notice which specifies the whole or part of those premises may be given under this section within the period of twelve months beginning with the date of the withdrawal or deemed withdrawal of the earlier notice or with the time when the order under section 23(1) becomes final (as the case may be).

(10) In subsections (8) and (9) any reference to a notice which specifies the whole or part of any premises includes a reference to a notice which specifies any premises which contain the whole or part of those premises; and in those subsections and this "specifies" means specifies under subsection (3)(a)(i).

(11) Where a notice is given in accordance with this section, then for the purposes of this Chapter the notice continues in force as from the relevant date—

(a) until a binding contract is entered into in pursuance of the notice, or an order is made under section 24(4)(a) or (b) or 25(6)(a) or (b) providing for the vesting of interests in the nominee purchaser;
(b) if the notice is withdrawn or deemed to have been withdrawn under or by virtue of any provision of this Chapter or under section 74(3), until the date of the withdrawal or deemed withdrawal, or
(c) until such other time as the notice ceases to have effect by virtue of any provision of this Chapter.

(12) In this Chapter "the specified premises", in relation to a claim made under this Chapter, means—

(a) the premises specified in the initial notice under subsection (3)(a)(i), or
(b) if it is subsequently agreed or determined under this Chapter that any less extensive premises should be acquired in pursuance of the notice in satisfaction of the claim, those premises;

and similarly references to any property or interest specified in the initial notice under subsection (3)(a)(ii) or (c)(i) shall, if it is subsequently agreed or determined under this Chapter that any less extensive property or interest should be acquired in pursuance of the notice, be read as references to that property or interest.

(13) Schedule 3 to this Act (which contains restrictions on participating in the exercise of the right to collective enfranchisement, and makes further provision in connection with the giving of notices under this section) shall have effect.

## Participating tenants and nominee purchaser

### 14 The participating tenants

(1) In relation to any claim to exercise the right to collective enfranchisement, the participating tenants are (subject to the provisions of this section and Part I of Schedule 3) the following persons, namely—

(a) in relation to the relevant date, the qualifying tenants by whom the initial notice is given; and
(b) in relation to any time falling after that date, such of those qualifying tenants as for the time being remain qualifying tenants of flats contained in the specified premises.

(2) Where the lease by virtue of which a participating tenant is a qualifying tenant of his flat is assigned to another person, the assignee of the lease shall, within the period of 14 days beginning with the date of the assignment, notify the nominee purchaser—

(a) of the assignment, and
(b) as to whether or not the assignee is electing to participate in the proposed acquisition.

(3) Where a qualifying tenant of a flat contained in the specified premises—

(a) is not one of the persons by whom the initial notice was given, and
(b) is not such an assignee of the lease of a participating tenant as is mentioned in subsection (2),

then (subject to paragraph 8 of Schedule 3) he may elect to participate in the proposed acquisition, but only with the agreement of all the persons who are for the time being participating tenants; and, if he does so elect, he shall notify the nominee purchaser forthwith of his election.

(4) Where a person notifies the nominee purchaser under subsection (2) or (3) of his election to participate in the proposed acquisition, he shall be regarded as a participating tenant for the purposes of this Chapter—

(a) as from the date of the assignment or agreement referred to in that subsection; and
(b) so long as he remains a qualifying tenant of a flat contained in the specified premises.

(5) Where a participating tenant dies, his personal representatives shall, within the period of 56 days beginning with the date of death, notify the nominee purchaser—

(a) of the death of the tenant, and
(b) as to whether or not the personal representatives are electing to withdraw from participation in the proposed acquisition;

and, unless the personal representatives of a participating tenant so notify the nominee purchaser that they are electing to withdraw from participation in that acquisition, they shall be regarded as a participating tenant for the purposes of this Chapter—
(i) as from the date of the death of the tenant, and
(ii) so long as his lease remains vested in them.

(6) Where in accordance with subsection (4) or (5) any assignee or personal representatives of a participating tenant ("the tenant") is or are to be regarded as a participating tenant for the purposes of this Chapter, any arrangements made between the nominee purchaser and the participating tenants and having effect immediately before the date of the assignment or (as the case may be) the date of death shall have effect as from that date—

*Appendix*

(a) with such modifications as are necessary for substituting the assignee or (as the case may be) the personal representatives as a party to the arrangements in the place of the tenant; or
(b) in the case of an assignment by a person who remains a qualifying tenant of a flat contained in the specified premises, with such modifications as are necessary for adding the assignee as a party to the arrangements.

(7) Where the nominee purchaser receives a notification under subsection (2), (3) or (5), he shall, within the period of 28 days beginning with the date of receipt of the notification—

(a) give a notice under subsection (8) to the reversioner in respect of the specified premises, and
(b) give a copy of that notice to every other relevant landlord.

(8) A notice under this subsection is a notice stating—

(a) in the case of a notification under subsection (2)—
   (i) the date of the assignment and the name and address of the assignee,
   (ii) that the assignee has or (as the case may be) has not become a participating tenant in accordance with subsection (4), and
   (iii) if he has become a participating tenant (otherwise than in a case to which subsection (6)(b) applies), that he has become such a tenant in place of his assignor;
(b) in the case of a notification under subsection (3), the name and address of the person who has become a participating tenant in accordance with subsection (4); and
(c) in the case of a notification under subsection (5)—
   (i) the date of death of the deceased tenant,
   (ii) the names and addresses of the personal representatives of the tenant, and
   (iii) that in accordance with that subsection those persons are or (as the case may be) are not to be regarded as a participating tenant.

(9) Every notice under subsection (8)—

(a) shall identify the flat with respect to which it is given; and

(b) if it states that any person or persons is or are to be regarded as a participating tenant, shall be signed by the person or persons in question.

(10) In this section references to assignment include an assent by personal representatives and assignment by operation of law, where the assignment is—

(a) to a trustee in bankruptcy, or
(b) to a mortgagee under section 89(2) of the Law of Property Act 1925 (foreclosure of leasehold mortgage),

and references to an assignee shall be construed accordingly.

(11) Nothing in this section has effect for requiring or authorising anything to be done at any time after a binding contract is entered into in pursuance of the initial notice.

## 15 The nominee purchaser: appointment and replacement

(1) The nominee purchaser shall conduct on behalf of the participating tenants all proceedings arising out of the initial notice, with a view to the eventual acquisition by him, on their behalf, of such freehold and other interests as fall to be so acquired under a contract entered into in pursuance of that notice.

(2) In relation to any claim to exercise the right to collective enfranchisement with respect to any premises, the nominee purchaser shall be such person or persons as may for the time being be appointed for the purposes of this section by the participating tenants; and in the first instance the nominee purchaser shall be the person or persons specified in the initial notice in pursuance of section 13(3)(f).

(3) The appointment of any person as the nominee purchaser, or as one of the persons constituting the nominee purchaser, may be terminated by the participating tenants by the giving of a notice stating that that person's appointment is to terminate on the date on which the notice is given.

(4) Any such notice must be given—

(a) to the person whose appointment is being terminated, and

*Appendix*

(b) to the reversioner in respect of the specified premises.

(5) Any such notice must in addition either—

(a) specify the name or names of the person or persons constituting the nominee purchaser as from the date of the giving of the notice, and an address in England and Wales at which notices may be given to that person or those persons under this Chapter; or
(b) state that the following particulars will be contained in a further notice given to the reversioner within the period of 28 days beginning with that date, namely—
   (i) the name of the person or persons for the time being constituting the nominee purchaser,
   (ii) if falling after that date, the date of appointment of that person or of each of those persons, and
   (iii) an address in England and Wales at which notices may be given to that person or those persons under this Chapter;

and the appointment of any person by way of replacement for the person whose appointment is being terminated shall not be valid unless his name is specified, or is one of those specified, under paragraph (a) or (b).

(6) Where the appointment of any person is terminated in accordance with this section, anything done by or in relation to the nominee purchaser before the date of termination of that person's appointment shall be treated, so far as necessary for the purpose of continuing its effect, as having been done by or in relation to the nominee purchaser as constituted on or after that date.

(7) Where the appointment of any person is so terminated, he shall not be liable under section 33 for any costs incurred in connection with the proposed acquisition under this Chapter at any time after the date of termination of his appointment; but if—

(a) at any such time he is requested by the nominee purchaser for the time being to supply to the nominee purchaser, at an address in England and Wales specified in the request, all or any documents in his custody or under his control that relate to that acquisition, and
(b) he fails without reasonable cause to comply with any such request or is guilty of any unreasonable delay in complying with it,

he shall be liable for any costs which are incurred by the nominee purchaser, or for which the nominee purchaser is liable under section 33, in consequence of the failure.

(8) Where—

(a) two or more persons together constitute the nominee purchaser, and
(b) the appointment of any (but not both or all) of them is terminated in accordance with this section without any person being appointed by way of immediate replacement,

the person or persons remaining shall for the time being constitute the nominee purchaser.

(9) Where—

(a) a notice given under subsection (3) contains such a statement as is mentioned in subsection (5)(b), and
(b) as a result of the termination of the appointment in question there is no nominee purchaser for the time being,

the running of any period which—
  (i) is prescribed by or under this Part for the giving of any other notice or the making of any application, and
  (ii) would otherwise expire during the period beginning with the date of the giving of the notice under subsection (3) and ending with the date when the particulars specified in subsection (5)(b) are notified to the reversioner,

shall (subject to subsection (10)) be suspended throughout the period mentioned in paragraph (ii).

(10) If—

(a) the circumstances are as mentioned in subsection (9)(a) and (b), but
(b) the particulars specified in subsection (5)(b) are not notified to the reversioner within the period of 28 days specified in that provision,

the initial notice shall be deemed to have been withdrawn at the end of that period.

(11) A copy of any notice given under subsection (3) or (5)(b) shall be given by the participating tenants to every relevant landlord (other

than the reversioner) to whom the initial notice or a copy of it was given in accordance with section 13 and Part II of Schedule 3; and, where a notice under subsection (3) terminates the appointment of a person who is one of two or more persons together constituting the nominee purchaser, a copy of the notice shall also be so given to every other person included among those persons.

(12) Nothing in this section applies in relation to the termination of the appointment of the nominee purchaser (or of any of the persons constituting the nominee purchaser) at any time after a binding contract is entered into in pursuance of the initial notice; and in this Chapter references to the nominee purchaser, so far as referring to anything done by or in relation to the nominee purchaser at any time falling after such a contract is so entered into, are references to the person or persons constituting the nominee purchaser at the time when the contract is entered into or such other person as is for the time being the purchaser under the contract.

## 16 The nominee purchaser: retirement or death

(1) The appointment of any person as the nominee purchaser, or as one of the persons constituting the nominee purchaser, may be terminated by that person by the giving of a notice stating that he is resigning his appointment with effect from 21 days after the date of the notice.

(2) Any such notice must be given—

(a) to each of the participating tenants; and
(b) to the reversioner in respect of the specified premises.

(3) Where the participating tenants have received any such notice, they shall, within the period of 56 days beginning with the date of the notice, give to the reversioner a notice informing him of the resignation and containing the following particulars, namely—

(a) the name or names of the person or persons for the time being constituting the nominee purchaser,
(b) if falling after that date, the date of appointment of that person or of each of those persons, and
(c) an address in England and Wales at which notices may be given to that person or those persons under this Chapter;

and the appointment of any person by way of replacement for the person resigning his appointment shall not be valid unless his name is specified, or is one of those specified, under paragraph (a).

(4) Subsections (6) to (8) of section 15 shall have effect in connection with a person's resignation of his appointment in accordance with this section as they have effect in connection with the termination of a person's appointment in accordance with that section.

(5) Where the person, or one of the persons, constituting the nominee purchaser dies, the participating tenants shall, within the period of 56 days beginning with the date of death, give to the reversioner a notice informing him of the death and containing the following particulars, namely—

(a) the name or names of the person or persons for the time being constituting the nominee purchaser,
(b) if falling after that date, the date of appointment of that person or of each of those persons, and
(c) an address in England and Wales at which notices may be given to that person or those persons under this Chapter;

and the appointment of any person by way of replacement for the person who has died shall not be valid unless his name is specified, or is one of those specified, under paragraph (a).

(6) Subsections (6) and (8) of section 15 shall have effect in connection with the death of any such person as they have effect in connection with the termination of a person's appointment in accordance with that section.

(7) If—

(a) the participating tenants are required to give a notice under subsection (3) or (5), and
(b) as a result of the resignation or death referred to in that subsection there is no nominee purchaser for the time being,

the running of any period which—
(i) is prescribed by or under this Part for the giving of any other notice or the making of any application, and
(ii) would otherwise expire during the period beginning with the relevant date and ending with the date when the particulars

specified in that subsection are notified to the reversioner,

shall (subject to subsection (8)) be suspended throughout the period mentioned in paragraph (ii); and for this purpose "the relevant date" means the date of the notice of resignation under subsection (1) or the date of death (as the case may be).

(8) If—

(a) the circumstances are as mentioned in subsection (7)(a) and (b), but
(b) the participating tenants fail to give a notice under subsection (3) or (as the case may be) subsection (5) within the period of 56 days specified in that subsection,

the initial notice shall be deemed to have been withdrawn at the end of that period.

(9) Where a notice under subsection (1) is given by a person who is one of two or more persons together constituting the nominee purchaser, a copy of the notice shall be given by him to every other person included among those persons; and a copy of any notice given under subsection (3) or (5) shall be given by the participating tenants to every relevant landlord (other than the reversioner) to whom the initial notice or a copy of it was given in accordance with section 13 and Part II of Schedule 3.

(10) Nothing in this section applies in relation to the resignation or death of the nominee purchaser (or any of the persons together constituting the nominee purchaser) at any time after a binding contract is entered into in pursuance of the initial notice.

## Procedure following giving of initial notice

### 17 Rights of access
(1) Once the initial notice or a copy of it has been given in accordance with section 13 or Part II of Schedule 3 to the reversioner or to any other relevant landlord, that person and any person authorised to act on his behalf shall, in the case of—

(a) any part of the specified premises, or
(b) any part of any property specified in the notice under section 13(3)(a)(ii),

in which he has a freehold or leasehold interest which is included in the proposed acquisition by the nominee purchaser, have a right of access thereto for the purpose of enabling him to obtain a valuation of that interest in connection with the notice or if it is reasonable in connection with any other matter arising out of the claim to exercise the right to collective enfranchisement.

(2) Once the initial notice has been given in accordance with section 13, the nominee purchaser and any person authorised to act on his behalf shall have a right of access to—

(a) any part of the specified premises, or
(b) any part of any property specified in the notice under section 13(3)(a)(ii),

where such access is reasonably required by the nominee purchaser in connection with any matter arising out of the notice.

(3) A right of access conferred by this section shall be exercisable at any reasonable time and on giving not less than 10 days' notice to the occupier of any premises to which access is sought (or, if those premises are unoccupied, to the person entitled to occupy them).

## 18 Duty of nominee purchaser to disclose existence of agreements affecting specified premises etc

(1) If at any time during the period beginning with the relevant date and ending with the time when a binding contract is entered into in pursuance of the initial notice—

(a) there subsists between the nominee purchaser and a person other than a participating tenant any agreement (of whatever nature) providing for the disposal of a relevant interest, or
(b) if the nominee purchaser is a company, any person other than a participating tenant holds any share in that company by virtue of which a relevant interest may be acquired,

the existence of that agreement or shareholding shall be notified to the reversioner by the nominee purchaser as soon as possible after the agreement or shareholding is made or established or, if in existence on the relevant date, as soon as possible after that date.

(2) If—

(a) the nominee purchaser is required to give any notification under subsection (1) but fails to do so before the price payable to the reversioner or any other relevant landlord in respect of the acquisition of any interest of his by the nominee purchaser is determined for the purposes of Schedule 6, and

(b) it may reasonably be assumed that, had the nominee purchaser given the notification, it would have resulted in the price so determined being increased by an amount referable to the existence of any agreement or shareholding falling within subsection (1)(a) or (b),

the nominee purchaser and the participating tenants shall be jointly and severally liable to pay the amount to the reversioner or (as the case may be) the other relevant landlord.

(3) In subsection (1) "relevant interest" means any interest in, or in any part of the specified premises or any property specified in the initial notice under section 13(3)(a)(ii).

(4) Paragraph (a) of subsection (1) does not, however, apply to an agreement if the only disposal of such an interest for which it provides is one consisting in the creation of an interest by way of security for a loan.

## 19 Effect of initial notice as respects subsequent transactions by freeholder etc

(1) Where the initial notice has been registered in accordance with section 97(1), then so long as it continues in force—

(a) any person who owns the freehold of the whole or any part of the specified premises or the freehold of any property specified in the notice under section 13(3)(a)(ii) shall not—
   (i) make any disposal severing his interest in those premises or in that property, or
   (ii) grant out of that interest any lease under which, if it had been granted before the relevant date, the interest of the tenant

would to any extent have been liable on that date to acquisition by virtue of section 2(1)(a) or (b); and

(b) no other relevant landlord shall grant out of his interest in the specified premises or in any property so specified any such lease as is mentioned in paragraph (a)(ii);

and any transaction shall be void to the extent that it purports to effect any such disposal or any such grant of a lease as is mentioned in paragraph (a) or (b).

(2) Where the initial notice has been so registered and at any time when it continues in force—

(a) any person who owns the freehold of the whole or any part of the specified premises or the freehold of any property specified in the notice under section 13(3)(a)(ii) disposes of his interest in those premises or that property, or

(b) any other relevant landlord disposes of any interest of his specified in the notice under section 13(3)(c)(i),

subsection (3) below shall apply in relation to that disposal.

(3) Where this subsection applies in relation to any such disposal as is mentioned in subsection (2)(a) or (b), all parties shall for the purposes of this Chapter be in the same position as if the person acquiring the interest under the disposal—

(a) had become its owner before the initial notice was given (and was accordingly a relevant landlord in place of the person making the disposal), and

(b) had been given any notice or copy of a notice given under this Chapter to that person, and

(c) had taken all steps which that person had taken;

and, if any subsequent disposal of that interest takes place at any time when the initial notice continues in force, this subsection shall apply in relation to that disposal as if any reference to the person making the disposal included any predecessor in title of his.

(4) Where immediately before the relevant date there is in force a binding contract relating to the disposal to any extent—

(a) by any person who owns the freehold of the whole or any part of the specified premises or the freehold of any property specified in the notice under section 13(3)(a)(ii), or

(b) by any other relevant landlord,

of any interest of his falling within subsection (2)(a) or (b), then, so long as the initial notice continues in force, the operation of the contract shall be suspended so far as it relates to any such disposal.

(5) Where—

(a) the operation of a contract has been suspended under subsection (4) ("the suspended contract"), and

(b) a binding contract is entered into in pursuance of the initial notice,

then (without prejudice to the general law as to the frustration of contracts) the person referred to in paragraph (a) or (b) of that subsection shall, together with all other persons, be discharged from the further performance of the suspended contract so far as it relates to any such disposal as is mentioned in subsection (4).

(6) In subsections (4) and (5) any reference to a contract (except in the context of such a contract as is mentioned in subsection (5)(b)) includes a contract made in pursuance of an order of any court; but those subsections do not apply to any contract providing for the eventuality of a notice being given under section 13 in relation to the whole or part of the property in which any such interest as is referred to in subsection (4) subsists.

## 20 Right of reversioner to require evidence of tenant's right to participate

(1) The reversioner in respect of the specified premises may, within the period of 21 days beginning with the relevant date, give the nominee purchaser a notice requiring him, in the case of any person by whom the initial notice was given, to deduce the title of that person to the lease by virtue of which it is claimed that he is a qualifying tenant of a flat contained in the specified premises.

(2) The nominee purchaser shall comply with any such requirement within the period of 21 days beginning with the date of the giving of the notice.

(3) Where—

(a) the nominee purchaser fails to comply with a requirement under subsection (1) in the case of any person within the period mentioned in subsection (2), and
(b) the initial notice would not have been given in accordance with section 13(2)(b) if—
   (i) that person, and
   (ii) any other person in the case of whom a like failure by the nominee purchaser has occurred,
had been neither included among the persons who gave the notice nor included among the qualifying tenants of the flats referred to in that provision,

the initial notice shall be deemed to have been withdrawn at the end of that period.

## 21 Reversioner's counter-notice

(1) The reversioner in respect of the specified premises shall give a counter-notice under this section to the nominee purchaser by the date specified in the initial notice in pursuance of section 13(3)(g).

(2) The counter-notice must comply with one of the following requirements, namely—

(a) state that the reversioner admits that the participating tenants were on the relevant date entitled to exercise the right to collective enfranchisement in relation to the specified premises;
(b) state that, for such reasons as are specified in the counter-notice, the reversioner does not admit that the participating tenants were so entitled;
(c) contain such a statement as is mentioned in paragraph (a) or (b) above but state that an application for an order under subsection (1) of section 23 is to be made by such appropriate landlord (within the meaning of that section) as is specified in the counter-notice, on the grounds that he intends to redevelop the whole or a substantial part of the specified premises.

(3) If the counter-notice complies with the requirement set out in subsection (2)(a), it must in addition—

*Appendix*

(a) state which (if any) of the proposals contained in the initial notice are accepted by the reversioner and which (if any) of those proposals are not so accepted, and specify—
   (i) in relation to any proposal which is not so accepted, the reversioner's counter-proposal, and
   (ii) any additional leaseback proposals by the reversioner;
(b) if (in a case where any property specified in the initial notice under section 13(3)(a)(ii) is property falling within section 1(3)(b)) any such counter-proposal relates to the grant of rights or the disposal of any freehold interest in pursuance of section 1(4), specify—
   (i) the nature of those rights and the property over which it is proposed to grant them, or
   (ii) the property in respect of which it is proposed to dispose of any such interest,

   as the case may be;
(c) state which interests (if any) the nominee purchaser is to be required to acquire in accordance with subsection (4) below;
(d) state which rights (if any) any relevant landlord, desires to retain—
   (i) over any property in which he has any interest which is included in the proposed acquisition by the nominee purchaser, or
   (ii) over any property in which he has any interest which the nominee purchaser is to be required to acquire in accordance with subsection (4) below,

   on the grounds that the rights are necessary for the proper management or maintenance of property in which he is to retain a freehold or leasehold interest; and
(e) include a description of any provisions which the reversioner or any other relevant landlord considers should be included in any conveyance to the nominee purchaser in accordance with section 34 and Schedule 7.

(4) The nominee purchaser may be required to acquire on behalf of the participating tenants the interest in any property of any relevant landlord, if the property—

(a) would for all practical purposes cease to be of use and benefit to him, or
(b) would cease to be capable of being reasonably managed or maintained by him,

in the event of his interest in the specified premises or (as the case may be) in any other property being acquired by the nominee purchaser under this Chapter.

(5) Where a counter-notice specifies any interest in pursuance of subsection (3)(c), the nominee purchaser or any person authorised to act on his behalf shall, in the case of any part of the property in which that interest subsists, have a right of access thereto for the purpose of enabling the nominee purchaser to obtain, in connection with the proposed acquisition by him a valuation of that interest; and subsection (3) of section 17 shall apply in relation to the exercise of that right as it applies in relation to the exercise of a right of access conferred by that section.

(6) Every counter-notice must specify an address in England and Wales at which notices may be given to the reversioner under this Chapter.

(7) The reference in subsection (3)(a)(ii) to additional leaseback proposals is a reference to proposals which relate to the leasing back, in accordance with section 36 and Schedule 9, of flats or other units contained in the specified premises and which are made either—

(a) in respect of flats or other units in relation to which Part II of that Schedule is applicable but which were not specified in the initial notice under section 13(3)(c)(ii), or
(b) in respect of flats or other units in relation to which Part III of that Schedule is applicable.

(8) Schedule 4 (which imposes requirements as to the furnishing of information by the reversioner about the exercise of rights under Chapter II with respect to flats contained in the specified premises) shall have effect.

## Applications to court or leasehold valuation tribunal

### 22 Proceedings relating to validity of initial notice
(1) Where—

(a) the reversioner in respect of the specified premises has given the nominee purchaser a counter-notice under section 21 which (whether it complies with the requirement set out in subsection

(2)(b) or (c) of that section) contains such a statement as is mentioned in subsection (2)(b) of that section, but
(b) the court is satisfied, on an application made by the nominee purchaser, that the participating tenants were on the relevant date entitled to exercise the right to collective enfranchisement in relation to the specified premises,

the court shall by order make a declaration to that effect.

(2) Any application for an order under subsection (1) must be made not later than the end of the period of two months beginning with the date of the giving of the counter-notice to the nominee purchaser.

(3) If on any such application the court makes an order under subsection (1), then (subject to subsection (4)) the court shall make an order—

(a) declaring that the reversioner's counter-notice shall be of no effect, and
(b) requiring the reversioner to give a further counter-notice to the nominee purchaser by such date as is specified in the order.

(4) Subsection (3) shall not apply if—

(a) the counter-notice complies with the requirement set out in section 21(2)(c), and
(b) either—
   (i) an application for an order under section 23(1) is pending, or
   (ii) the period specified in section 23(3) as the period for the making of such an application has not expired.

(5) Subsections (3) to (5) of section 21 shall apply to any further counter-notice required to be given by the reversioner under subsection (3) above as if it were a counter-notice under that section complying with the requirement set out in subsection (2)(a) of that section.

(6) If an application by the nominee purchaser for an order under subsection (1) is dismissed by the court, the initial notice shall cease to have effect at the time when the order dismissing the application becomes final.

## 23 Tenants' claim liable to be defeated where landlord intends to redevelop

(1) Where the reversioner in respect of the specified premises has given a counter-notice under section 21 which complies with the requirement set out in subsection (2)(c) of that section, the court may, on the application of any appropriate landlord, by order declare that the right to collective enfranchisement shall not be exercisable in relation to those premises by reason of that landlord's intention to redevelop the whole or a substantial part of the premises.

(2) The court shall not make an order under subsection (1) unless it is satisfied—

(a) that not less than two-thirds of all the long leases on which flats contained in the specified premises are held are due to terminate within the period of five years beginning with the relevant date; and
(b) that for the purposes of redevelopment the applicant intends, once the leases in question have so terminated—

   (i) to demolish or reconstruct, or
   (ii) to carry out substantial works of construction on,

   the whole or a substantial part of the specified premises; and
(c) that he could not reasonably do so without obtaining possession of the flats demised by those leases.

(3) Any application for an order under subsection (1) must be made within the period of two months beginning with the date of the giving of the counter-notice to the nominee purchaser; but, where the counter-notice is one falling within section 22(1)(a), such an application shall not be proceeded with until such time (if any) as an order under section 22(1) becomes final.

(4) Where an order under subsection (1) is made by the court, the initial notice shall cease to have effect on the order becoming final.

(5) Where an application for an order under subsection (1) is dismissed by the court, the court shall make an order—

(a) declaring that the reversioner's counter-notice shall be of no effect, and
(b) requiring the reversioner to give a further counter-notice to the nominee purchaser by such date as is specified in the order.

(6) Where—

(a) the reversioner has given such a counter-notice as is mentioned in subsection (1), but
(b) either—
  (i) no application for an order under that subsection is made within the period referred to in subsection (3), or
  (ii) such an application is so made but is subsequently withdrawn,

then (subject to subsection (8)), the reversioner shall give a further counter-notice to the nominee purchaser within the period of two months beginning with the appropriate date.

(7) In subsection (6) "the appropriate date" means—

(a) if subsection (6)(b)(i) applies, the date immediately following the end of the period referred to in subsection (3); and
(b) if subsection (6)(b)(ii) applies, the date of withdrawal of the application.

(8) Subsection (6) shall not apply if any application has been made by the nominee purchaser under section 22(1).

(9) Subsections (3) to (5) of section 21 shall apply to any further counter-notice required to be given by the reversioner under subsection (5) or (6) above as if it were a counter-notice under that section complying with the requirement set out in subsection (2)(a) of that section.

(10) In this section "appropriate landlord", in relation to the specified premises, means—

(a) the reversioner or any other relevant landlord; or
(b) any two or more persons falling within paragraph (a) who are acting together.

## 24 Applications where terms in dispute or failure to enter contract

(1) Where the reversioner in respect of the specified premises has given the nominee purchaser —

(a) a counter-notice under section 21 complying with the requirement set out in subsection (2)(a) of that section, or

(b) a further counter-notice required by or by virtue of section 22(3) or section 23(5) or (6),

but any of the terms of acquisition remain in dispute at the end of the period of two months beginning with the date on which the counter-notice or further counter-notice was so given, a leasehold valuation tribunal may, on the application of either the nominee purchaser or the reversioner, determine the matters in dispute.

(2) Any application under subsection (1) must be made not later than the end of the period of six months beginning with the date on which the counter-notice or further counter-notice was given to the nominee purchaser.

(3) Where—

(a) the reversioner has given the nominee purchaser such a counter-notice or further counter-notice as is mentioned in subsection (1)(a) or (b), and
(b) all of the terms of acquisition have been either agreed between the parties or determined by a leasehold valuation tribunal under subsection (1),

but a binding contract incorporating those terms has not been entered into by the end of the appropriate period specified in subsection (6), the court may, on the application of either the nominee purchaser or the reversioner, make such order under subsection (4) as it thinks fit.

(4) The court may under this subsection make an order—

(a) providing for the interests to be acquired by the nominee purchaser to be vested in him on the terms referred to in subsection (3);
(b) providing for those interests to be vested in him on those terms, but subject to such modifications as—
  (i) may have been determined by a leasehold valuation tribunal, on the application of either the nominee purchaser or the reversioner, to be required by reason of any change in circumstances since the time when the terms were agreed or determined as mentioned in that subsection, and
  (ii) are specified in the order; or
(c) providing for the initial notice to be deemed to have been

withdrawn at the end of the appropriate period specified in subsection (6);

and Schedule 5 shall have effect in relation to any such order as is mentioned in paragraph (a) or (b) above.

(5) Any application for an order under subsection (4) must be made not later than the end of the period of two months beginning immediately after the end of the appropriate period specified in subsection (6).

(6) For the purposes of this section the appropriate period is—

(a) where all of the terms of acquisition have been agreed between the parties, the period of two months beginning with the date when those terms were finally so agreed;
(b) where all or any of those terms have been determined by a leasehold valuation tribunal under subsection (1)—
   (i) the period of two months beginning with the date when the decision of the tribunal under that subsection becomes final, or
   (ii) such other period as may have been fixed by the tribunal when making its determination.

(7) In this section "the parties" means the nominee purchaser and the reversioner and any relevant landlord who has given to those persons a notice for the purposes of paragraph 7(1)(a) of Schedule 1.

(8) In this Chapter "the terms of acquisition", in relation to a claim made under this Chapter, means the terms of the proposed acquisition by the nominee purchaser, whether relating to—

(a) the interests to be acquired,
(b) the extent of the property to which those interests relate or the rights to be granted over any property,
(c) the amounts payable as the purchase price for such interests,
(d) the apportionment of conditions or other matters in connection with the severance of any reversionary interest, or
(e) the provisions to be contained in any conveyance,

or otherwise, and includes any such terms in respect of any interest to be acquired in pursuance of section 1(4) or 21(4).

## 25 Applications where reversioner fails to give counter-notice or further counter-notice

(1) Where the initial notice has been given in accordance with section 13 but—

(a) the reversioner has failed to give the nominee purchaser a counter-notice in accordance with section 21(1), or
(b) if required to give the nominee purchaser a further counter-notice by or by virtue of section 22(3) or section 23(5) or (6), the reversioner has failed to comply with that requirement,

the court may, on the application of the nominee purchaser, make an order determining the terms on which he is to acquire, in accordance with the proposals contained in the initial notice, such interests and rights as are specified in it under section 13(3).

(2) The terms determined by the court under subsection (1) shall, if Part II of Schedule 9 is applicable, include terms which provide for the leasing back, in accordance with section 36 and that Part of that Schedule, of flats or other units contained in the specified premises.

(3) The court shall not make any order on an application made by virtue of paragraph (a) of subsection (1) unless it is satisfied—

(a) that the participating tenants were on the relevant date entitled to exercise the right to collective enfranchisement in relation to the specified premises; and
(b) if applicable, that the requirements of Part II of Schedule 3 were complied with as respects the giving of copies of the initial notice.

(4) Any application for an order under subsection (1) must be made not later than the end of the period of six months beginning with the date by which the counter-notice or further counter-notice referred to in that subsection was to be given to the nominee purchaser.

(5) Where—

(a) the terms of acquisition have been determined by an order of the court under subsection (1), but
(b) a binding contract incorporating those terms has not been entered into by the end of the appropriate period specified in subsection (8),

the court may, on the application of either the nominee purchaser or the reversioner, make such order under subsection (6) as it thinks fit.

(6) The court may under this subsection make an order—

(a) providing for the interests to be acquired by the nominee purchaser to be vested in him on the terms referred to in subsection (5);
(b) providing for those interests to be vested in him on those terms, but subject to such modifications as—
   (i) may have been determined by a leasehold valuation tribunal, on the application of either the nominee purchaser or the reversioner, to be required by reason of any change in circumstances since the time when the terms were determined as mentioned in that subsection, and
   (ii) are specified in the order; or
(c) providing for the initial notice to be deemed to have been withdrawn at the end of the appropriate-period specified in subsection (8);

and Schedule 5 shall have effect in relation to any such order as is mentioned in paragraph (a) or (b) above.

(7) Any application for an order under subsection (6) must be made not later than the end of the period of two months beginning immediately after the end of the appropriate period specified in subsection (8).

(8) For the purposes of this section the appropriate period is—

(a) the period of two months beginning with the date when the order of the court under subsection (1) becomes final, or
(b) such other period as may have been fixed by the court when making that order.

## 26 Applications where relevant landlord cannot be found

(1) Where not less than two-thirds of the qualifying tenants of flats contained in any premises to which this Chapter applies desire to make a claim to exercise the right to collective enfranchisement in relation to those premises but—

(a) (in a case to which section 9(1) applies) the person who owns the freehold of the premises cannot be found or his identity cannot be ascertained, or
(b) (in a case to which section 9(2) or (2A) applies) each of the relevant landlords is someone who cannot be found or whose identity cannot be ascertained,

the court may, on the application of the qualifying tenants in question, make a vesting order under this subsection—
(i) with respect to any interests of that person (whether in those premises or in any other property) which are liable to acquisition on behalf of those tenants by virtue of section 1(1) or (2)(a) or section 2(1), or
(ii) with respect to any interests of those landlords which are so liable to acquisition by virtue of any of those provisions,

as the case may be.

(2) Where in a case to which section 9(2) applies—

(a) not less than two-thirds of the qualifying tenants of flats contained in any premises to which this Chapter applies desire to make a claim to exercise the right to collective enfranchisement in relation to those premises, and
(b) paragraph (b) of subsection (1) does not apply, but
(c) a notice of that claim or (as the case may be) a copy of such a notice cannot be given in accordance with section 13 or Part II of Schedule 3 to any person to whom it would otherwise be required to be so given because he cannot be found or his identity cannot be ascertained,

the court may, on the application of the qualifying tenants in question, make an order dispensing with the need to give such a notice or (as the case may be) a copy of such a notice to that person.

(3) If, in a case to which section 9(2) applies, that person is the person who owns the freehold of the premises, then on the application of those tenants, the court may, in connection with an order under subsection (2), make an order appointing any other relevant landlord to be the reversioner in respect of the premises in place of that person; and if it does so references in this Chapter to the reversioner shall apply accordingly.

(3A) Where in a case to which section 9(2A) applies—

(a) not less than two-thirds of the qualifying tenants of flats contained in any premises to which this Chapter applies desire to make a claim to exercise the right to collective enfranchisement in relation to those premises, and

*Appendix*

(b) paragraph (b) of subsection (1) does not apply, but
(c) a copy of a notice of that claim cannot be given in accordance with Part II of Schedule 3 to any person to whom it would otherwise be required to be so given because he cannot be found or his identity cannot be ascertained,

the court may, on the application of the qualifying tenants in question, make an order dispensing with the need to give a copy of such a notice to that person.

(4) The court shall not make an order on any application under subsection (1), (2) or (3A) unless it is satisfied—

(a) that on the date of the making of the application the premises to which the application relates were premises to which this Chapter applies; and
(b) that on that date the applicants would not have been precluded by any provision of this Chapter from giving a valid notice under section 13 with respect to those premises

(5) Before making any such order the court may require the applicants to take such further steps by way of advertisement or otherwise as the court thinks proper for the purpose of tracing the person or persons in question; and if, after an application is made for a vesting order under subsection (1) and before any interest is vested in pursuance of the application, the person or (as the case may be) any of the persons referred to in paragraph (a) or (b) of that subsection is traced, then no further proceedings shall be taken with a view to any interest being so vested, but (subject to subsection (6))—

(a) the rights and obligations of all parties shall be determined as if the applicants had, at the date of the application, duly given notice under section 13 of their claim to exercise the right to collective enfranchisement in relation to the premises to which the application relates; and
(b) the court may give such directions as the court thinks fit as to the steps to be taken for giving effect to those rights and obligations, including directions modifying or dispensing with any of the requirements of this Chapter or of regulations made under this Part.

(6) An application for a vesting order under subsection (1) may be withdrawn at any time before execution of a conveyance under section

27(3) and, after it is withdrawn, subsection (5)(a) above shall not apply; but where any step is taken (whether by the applicants or otherwise) for the purpose of giving effect to subsection (5)(a) in the case of any application, the application shall not afterwards be withdrawn except—

(a) with the consent of every person who is the owner of any interest the vesting of which is sought by the applicants, or
(b) by leave of the court,

and the court shall not give leave unless it appears to the court just to do so by reason of matters coming to the knowledge of the applicants in consequence of the tracing of any such person.

(7) Where an order has been made under subsection (2) or (3A) dispensing with the need to give a notice under section 13, or a copy of such a notice, to a particular person with respect to any particular premises, then if—

(a) a notice is subsequently given under that section with respect to those premises, and
(b) in reliance on the order, the notice or a copy of the notice is not to be given to that person,

the notice must contain a statement of the effect of the order.

(8) Where a notice under section 13 contains such a statement in accordance with subsection (7) above, then in determining for the purposes of any provision of this Chapter whether the requirements of section 13 or Part II of Schedule 3 have been complied with in relation to the notice, those requirements shall be deemed to have been complied with so far as relating to the giving of the notice or a copy of it to the person referred to in subsection (7) above.

(9) Rules of court shall make provision—

(a) for requiring notice of any application under subsection (3) to be served by the persons making the application on any person who the applicants know or have reason to believe is a relevant landlord; and
(b) for enabling persons served with any such notice to be joined as parties to the proceedings.

## 27 Supplementary provisions relating to vesting orders under section 26(1)

(1) A vesting order under section 26(1) is an order providing for the vesting of any such interests as are referred to in paragraph (i) or (ii) of that provision—

(a) in such person or persons as may be appointed for the purpose by the applicants for the order, and
(b) on such terms as may be determined by a leasehold valuation tribunal to be appropriate with a view to the interests being vested in that person or those person in like manner (so far as the circumstances permit) as if the applicants had, at the date of their application, given notice under section 13 of their claim to exercise the right to collective enfranchisement in relation to the premises with respect to which the order is made.

(2) If a leasehold valuation tribunal so determines in the case of a vesting order under section 26(1), the order shall have effect in relation to interests which are less extensive than those specified in the application on which the order was made.

(3) Where any interests are to be vested in any person or person by virtue of a vesting order under section 26(1), then on his or their paying into court the appropriate sum in respect of each of those interests there shall be executed by such person as the court may designate a conveyance which—

(a) is in a form approved by a leasehold valuation tribunal, and
(b) contains such provisions as may be so approved for the purpose of giving effect so far as possible to the requirements of section 34 and Schedule 7;

and that conveyance shall be effective to vest in the person or persons to whom the conveyance is made the interests expressed to be conveyed, subject to and in accordance with the terms of the conveyance.

(4) In connection with the determination by a leasehold valuation tribunal of any question as to the interests to be conveyed by any such conveyance, or as to the rights with or subject to which they are to be conveyed, it shall be assumed (unless the contrary is shown) that any person whose interests are to be conveyed ("the transferor") has no interest in property other than those interests and, for the purpose of

excepting them from the conveyance, any minerals underlying the property in question.

(5) The appropriate sum which in accordance with subsection (3) is to be paid into court in respect of any interest is the aggregate of—

(a) such amount as may be determined by a leasehold valuation tribunal to be the price which would be payable in respect of that interest in accordance with Schedule 6 if the interest were being acquired in pursuance of such a notice as is mentioned in subsection (1)(b); and
(b) any amounts or estimated amounts determined by such a tribunal as being, at the time of execution of the conveyance, due to the transferor from any tenants of his of premises comprised in the premises in which that interest subsists (whether due under or in respect of their leases or under or in respect of agreements collateral thereto).

(6) Where any interest is vested in any person or persons in accordance with this section, the payment into court of the appropriate sum in respect of that interest shall be taken to have satisfied any claims against the applicants for the vesting order under section 26(1), their personal representatives or assigns in respect of the price payable under this Chapter for the acquisition of that interest.

(7) Where any interest is so vested in any person or persons, section 32(5) shall apply in relation to his or their acquisition of that interest as it applies in relation to the acquisition of any interest by a nominee purchaser.

## Termination of acquisition procedures

### 28 Withdrawal from acquisition by participating tenants
(1) At any time before a binding contract is entered into in pursuance of the initial notice, the participating tenants may withdraw that notice by the giving of a notice to that effect under this section ("a notice of withdrawal").

(2) A notice of withdrawal must be given—

*Appendix*

(a) to the nominee purchaser;
(b) to the reversioner in respect of the specified premises; and
(c) to every other relevant landlord who is known or believed by the participating tenants to have given to the nominee purchaser a notice under paragraph 7(1) or (4) of Schedule 1;

and, if by virtue of paragraph (c) a notice of withdrawal falls to be given to any person falling within that paragraph, it shall state that he is a recipient of the notice.

(3) The nominee purchaser shall, on receiving a notice of withdrawal, give a copy of it to every relevant landlord who—

(a) has given to the nominee purchaser such a notice as is mentioned in subsection (2)(c); and
(b) is not stated in the notice of withdrawal to be a recipient of it.

(4) Where a notice of withdrawal is given by the participating tenants under subsection (1)—

(a) those persons, and
(b) (subject to subsection (5)) every other person who is not a participating tenant for the time being but has at any time been such a tenant,

shall be liable—

(i) to the reversioner, and
(ii) to every other relevant landlord,

for all relevant costs incurred by him in pursuance of the initial notice down to the time when the notice of withdrawal or a copy of it is given to him in accordance with subsection (2) or (3).

(5) A person falling within paragraph (b) of subsection (4) shall not be liable for any costs by virtue of that subsection if—

(a) the lease in respect of which he was a participating tenant has been assigned to another person; and
(b) that other person has become a participating tenant in accordance with section 14(4);

and in paragraph (a) above the reference to an assignment shall be construed in accordance with section 14(10).

(6) Where any liability for costs arises under subsection (4)—

(a) it shall be a joint and several liability of the persons concerned; and
(b) the nominee purchaser shall not be liable for any costs under section 33.

(7) In subsection (4) "relevant costs", in relation to the reversioner or any other relevant landlord, means costs for which the nominee purchaser would (apart from subsection (6)) be liable to that person under section 33.

## 29 Deemed withdrawal of initial notice

(1) Where, in a case falling within paragraph (a) of subsection (1) of section 22—

(a) no application for an order under that subsection is made within the period specified in subsection (2) of that section, or
(b) such an application is so made but is subsequently withdrawn, the initial notice shall be deemed to have been withdrawn—
   (i) (if paragraph (a) above applies) at the end of that period, or
   (ii) (if paragraph (b) above applies) on the date of the withdrawal of the application.

(2) Where—

(a) in a case to which subsection (1) of section 24 applies, no application under that subsection is made within the period specified in subsection (2) of that section, or
(b) in a case to which subsection (3) of that section applies, no application for an order under subsection (4) of that section is made within the period specified in subsection (5) of that section,

the initial notice shall be deemed to have been withdrawn at the end of the period referred to in paragraph (a) or (b) above (as the case may be).

(3) Where, in a case falling within paragraph (a) or (b) of subsection (1) of section 25, no application for an order under that subsection is made within the period specified in subsection (4) of that section, the initial notice shall be deemed to have been withdrawn at the end of that period.

*Appendix*

(4) Where, in a case to which subsection (5) of section 25 applies, no application for an order under subsection (6) of that section is made within the period specified in subsection (7) of that section, the initial notice shall be deemed to have been withdrawn at the end of that period.

(5) The following provisions, namely—

(a) section 15(10),
(b) section 16(8),
(c) section 20(3),
(d) section 24(4)(c), and
(e) section 25(6)(c),

also make provision for a notice under section 13 to be deemed to have been withdrawn at a particular time.

(6) Where the initial notice is deemed to have been withdrawn at any time by virtue of any provision of this Chapter, subsections (4) and (5) of section 28 shall apply for the purposes of this section in like manner as they apply where a notice of withdrawal is given under that section, but as if the reference in subsection (4) of that section to the time when a notice or copy is given as there mentioned were a reference to the time when the initial notice is so deemed to have been withdrawn.

(7) Where the initial notice is deemed to have been withdrawn by virtue of section 15(10) or 16(8)—

(a) the liability for costs arising by virtue of subsection (6) above shall be a joint and several liability of the persons concerned; and
(b) the nominee purchaser shall not be liable for any costs under section 33.

(8) In the provisions applied by subsection (6), "relevant costs", in relation to the reversioner or any other relevant landlord, means costs for which the nominee purchaser is, or would (apart from subsection (7)) be, liable to that person under section 33.

## 30 Effect on initial notice or subsequent contract of institution of compulsory acquisition procedures

(1) A notice given under section 13 shall be of no effect if on the relevant date—

(a) any acquiring authority has, with a view to the acquisition of the whole or part of the specified premises for any authorised purpose—
   (i) served notice to treat on any relevant person, or
   (ii) entered into a contract for the purchase of the interest of any such person in the premises or part of them, and
(b) the notice to treat or contract remains in force.

(2) In subsection (1) "relevant person", in relation to the specified premises, means—

(a) the person who owns the freehold of the premises [or, where the freehold of the whole of the premises is not owned by the same person, any person who owns the freehold of part of them]; or
(b) any other person who owns any leasehold interest in the premises which is specified in the initial notice under section 13(3)(c)(i).

(3) A notice given under section 13 shall not specify under subsection (3)(a)(ii) or (c)(i) of that section any property or leasehold interest in property if on the relevant date—

(a) any acquiring authority has, with a view to the acquisition of the whole or part of the property for any authorised purpose—
   (i) served notice to treat on the person who owns the freehold of, or any such leasehold interest in, the property, or
   (ii) entered into a contract for the purchase of the interest of any such person in the property or part of it, and
(b) the notice to treat or contract remains in force.

(4) A notice given under section 13 shall cease to have effect if before a binding contract is entered into in pursuance of the notice, any acquiring authority serves, with a view to the acquisition of the whole or part of the specified premises for any authorised purpose, notice to treat as mentioned in subsection (1)(a).

(5) Where any such authority so serves notice to treat at any time after a binding contract is entered into in pursuance of the notice given under section 13 but before completion of the acquisition by the nominee purchaser under this Chapter, then (without prejudice to the general law as to the frustration of contracts) the parties to the contract shall be discharged from the further performance of the contract.

(6) Where subsection (4) or (5) applies in relation to the initial notice or any contract entered into in pursuance of it, then on the occasion of the compulsory acquisition in question the compensation payable in respect of any interest in the specified premises (whether or not the one to which the relevant notice to treat relates) shall be determined on the basis of the value of the interest—

(a) (if subsection (4) applies) subject to and with the benefit of the rights and obligations arising from the initial notice and affecting that interest; or
(b) (if subsection (5) applies) subject to and with the benefit of the rights and obligations arising from the contract and affecting that interest.

(7) In this section—

(a) "acquiring authority", in relation to the specified premises or any other property, means any person or body of persons who has or have been, or could be, authorised to acquire the whole or part of those premises or that property compulsorily for any purpose; and
(b) "authorised purpose", in relation to any acquiring authority, means any such purpose.

## 31 Effect on initial notice of designation for inheritance tax purposes and applications for designation

(1) A notice given under section 13 shall be of no effect if on the relevant date the whole or any part of—

(a) the specified premises, or
(b) any property specified in the notice under section 13(3)(a)(ii), is qualifying property.

(2) For the purposes of this section the whole or any part of the specified premises, or of any property specified as mentioned in subsection (1), is qualifying property if—

(a) it has been designated under section 31(1)(b), (c) or (d) of the Inheritance Tax Act 1984 (designation and undertakings relating to conditionally exempt transfers), whether with or without any other property, and no chargeable event has subsequently occurred with respect to it; or

(b) an application to the Board for it to be so designated is pending; or
(c) it is the property of a body not established or conducted for profit and a direction has been given in relation to it under section 26 of that Act (gifts for public benefit), whether with or without any other property; or
(d) an application to the Board for a direction to be so given in relation to it is pending.

(3) For the purposes of subsection (2) an application is pending as from the time when it is made to the Board until such time as it is either granted or refused by the Board or withdrawn by the applicant; and for this purpose an application shall not be regarded as made unless and until the applicant has submitted to the Board all such information in support of the application as is required by the Board.

(4) A notice given under section 13 shall cease to have effect if, before a binding contract is entered into in pursuance of the notice, the whole or any part of—

(a) the specified premises, or
(b) any property specified in the notice under section 13(3)(a)(ii), becomes qualifying property.

(5) Where a notice under section 13 ceases to have effect by virtue of subsection (4) above—

(a) the nominee purchaser shall not be liable for any costs under section 33; and
(b) the person who applied or is applying for designation or a direction shall be liable—
   (i) to the qualifying tenants by whom the notice was given for all reasonable costs incurred by them in the preparation and giving of the notice; and
   (ii) to the nominee purchaser for all reasonable costs incurred in pursuance of the notice by him or by any other person who has acted as the nominee purchaser.

(6) Where it is claimed that subsection (1) or (4) applies in relation to a notice under section 13, the person making the claim shall, at the time of making it, furnish the nominee purchaser with evidence in support

of it; and if he fails to do so he shall be liable for any costs which are reasonably incurred by the nominee purchaser in consequence of the failure.

(7) In subsection (2)—

(a) paragraphs (a) and (b) apply to designation under section 34(1)(a), (b) or (c) of the Finance Act 1975 or section 77(1)(b), (c) or (d) of the Finance Act 1976 as they apply to designation under section 31(1)(b), (c) or (d) of the Inheritance Tax Act 1984; and
(b) paragraphs (c) and (d) apply to a direction under paragraph 13 of Schedule 6 to the Finance Act 1975 as they apply to a direction under section 26 of that Act of 1984.

(8) In this section—
"the Board" means the Commissioners of Inland Revenue;
"chargeable event" means—

(a) any event which in accordance with any provision of Chapter II of Part II of the Inheritance Tax Act 1984 (exempt transfers) is a chargeable event, including any such provision as applied by section 78(3) of that Act (conditionally exempt occasions); or
(b) any event which would have been a chargeable event in the circumstances mentioned in section 79(3) of that Act (exemption from ten-yearly charge).

## Determination of price and costs of enfranchisement

### 32 Determination of price

(1) Schedule 6 to this Act (which relates to the determination of the price payable by the nominee purchaser in respect of each of the freehold and other interests to be acquired by him in pursuance of this Chapter) shall have effect.

(2) The lien of the owner of any such interest (as vendor) on the specified premises, or (as the case may be) on any other property, for the price payable shall extend—

(a) to any amounts which, at the time of the conveyance of that interest, are due to him from any tenants of his of premises

comprised in the premises in which that interest subsists (whether due under or in respect of their leases or under or in respect of agreements collateral thereto); and
(b) to any amount payable to him by virtue of section 18(2); and
(c) to any costs payable to him by virtue of section 33.

(3) Subsection (2)(a) does not apply in relation to amounts due to the owner of any such interest from tenants of any premises which are to be comprised in the premises demised by a lease granted in accordance with section 36 and Schedule 9.

(4) In subsection (2) the reference to the specified premises or any other property includes a reference to a part of those premises or that property.

(5) Despite the fact that in accordance with Schedule 6 no payment or only a nominal payment is payable by the nominee purchaser in respect of the acquisition by him of any interest he shall nevertheless be deemed for all purposes to be a purchaser of that interest for a valuable consideration in money or money's worth.

**33 Costs of enfranchisement**
(1) Where a notice is given under section 13, then (subject to the provisions of this section and sections 28(6), 29(7) and 31(5)) the nominee purchaser shall be liable, to the extent that they have been incurred in pursuance of the notice by the reversioner or by any other relevant landlord, for the reasonable costs of and incidental to any of the following matters, namely—

(a) any investigation reasonably undertaken—
  (i) of the question whether any interest in the specified premises or other property is liable to acquisition in pursuance of the initial notice, or
  (ii) of any other question arising out of that notice;
(b) deducing, evidencing and verifying the title to any such interest;
(c) making out and furnishing such abstracts and copies as the nominee purchaser may require;
(d) any valuation of any interest in the specified premises or other property;
(e) any conveyance of any such interest;

but this subsection shall not apply to any costs if on a sale made voluntarily a stipulation that they were to be borne by the purchaser would be void.

(2)   For the purposes of subsection (1) any costs incurred by the reversioner or any other relevant landlord in respect of professional services rendered by any person shall only be regarded as reasonable if and to the extent that costs in respect of such services might reasonably be expected to have been incurred by him if the circumstances had been such that he was personally liable for all such costs.

(3)   Where by virtue of any provision of this Chapter the initial notice ceases to have effect at any time, then (subject to subsection (4)) the nominee purchaser's liability under this section for costs incurred by any person shall be a liability for costs incurred by him down to that time.

(4)   The nominee purchaser shall not be liable for any costs under this section if the initial notice ceases to have effect by virtue of section 23(4) or 30(4).

(5)   The nominee purchaser shall not be liable under this section for any costs which a party to any proceedings under this Chapter before a leasehold valuation tribunal incurs in connection with the proceedings.

(6)   In this section references to the nominee purchaser include references to any person whose appointment has terminated in accordance with section 15(3) or 16(1); but this section shall have effect in relation to such a person subject to section 15(7).

(7)   Where by virtue of this section, or of this section and section 29(6) taken together, two or more persons are liable for any costs, they shall be jointly and severally liable for them.

## Completion of acquisition

### 34 Conveyance to nominee purchaser
(1)   Any conveyance executed for the purposes of this Chapter, being a conveyance to the nominee purchaser of the freehold of the specified

premises, of a part of those premises or of any other property, shall grant to the nominee purchaser an estate in fee simple absolute in those premises[, that part of those premises] or that property, subject only to such incumbrances as may have been agreed or determined under this Chapter to be incumbrances subject to which that estate should be granted, having regard to the following provisions of this Chapter.

(2) Any such conveyance shall, where the nominee purchaser is to acquire any leasehold interest in the specified premises, the part of the specified premises or (as the case may be) in the other property to which the conveyance relates, provide for the disposal to the nominee purchaser of any such interest.

(3) Any conveyance executed for the purposes of this Chapter shall have effect under section 2(1) of the Law of Property Act 1925 (conveyances overreaching certain equitable interests etc) to overreach any incumbrance capable of being overreached under section 2(1)—

(a) as if, where the interest conveyed is settled land for the purposes of the Settled Land Act 1925, the conveyance were made under the powers of that Act, and
(b) as if the requirements of section 2(1) as to payment of the capital money allowed any part of the purchase price paid or applied in accordance with section 35 below or Schedule 8 to this Act to be so paid or applied.

(4) For the purposes of this section "incumbrances" includes—

(a) rentcharges, and
(b) (subject to subsection (5)) personal liabilities attaching in respect of the ownership of land or an interest in land though not charged on that land or interest.

(5) Burdens originating in tenure, and burdens in respect of the upkeep or regulation for the benefit of any locality of any land, building, structure, works, ways or watercourse shall not be treated as incumbrances for the purposes of this section; but any conveyance executed for the purposes of this Chapter shall be made subject to any such burdens.

(6) A conveyance executed for the purposes of this Chapter shall not be made subject to any incumbrance capable of being overreached by

*Appendix*

the conveyance, but shall be made subject (where they are not capable of being overreached) to—

(a) rentcharges redeemable under sections 8 to 10 of the Rentcharges Act 1977, and
(b) those falling within paragraphs (c) and (d) of section 2(3) of that Act (estate rentcharges and rentcharges imposed under certain enactments),

except as otherwise provided by subsections (7) and (8) below.

(7) Where any land is to be conveyed to the nominee purchaser by a conveyance executed for the purposes of this Chapter, subsection (6) shall not preclude the person who owns the freehold interest in the land from releasing, or procuring the release of, the land from any rentcharge.

(8) The conveyance of any such land ("the relevant land") may, with the agreement of the nominee purchaser (which shall not be unreasonably withheld), provide in accordance with section 190(1) of the Law of Property Act 1925 (charging of rentcharges on land without rent owner's consent) that a rentcharge—

(a) shall be charged exclusively on other land affected by it in exoneration of the relevant land, or
(b) shall be apportioned between other land affected by it and the relevant land.

(9) Except to the extent that any departure is agreed to by the nominee purchaser and the person whose interest is to be conveyed, any conveyance executed for the purposes of this Chapter shall—

(a) as respects the conveyance of any freehold interest, conform with the provisions of Schedule 7, and
(b) as respects the conveyance of any leasehold interest, conform with the provisions of paragraph 2 of that Schedule (any reference in that paragraph to the freeholder being read as a reference to the person whose leasehold interest is to be conveyed), and with the reference to the covenants for title implied under Part I of the Law of Property (Miscellaneous Provisions) Act 1994 being read as excluding the covenant in section 4(1)(b) of that Act (compliance with terms of lease)].

(10) Any such conveyance shall in addition contain a statement that it

is a conveyance executed for the purposes of this Chapter; and any such statement shall comply with such requirements as may be prescribed by land registration rules under the Land Registration Act 2002.

## 35 Discharge of existing mortgages on transfer to nominee purchaser [RTE company]

(1) Subject to the provisions of Schedule 8, where any interest is acquired by the nominee purchaser in pursuance of this Chapter, the conveyance by virtue of which it is so acquired shall, as regards any mortgage to which this section applies, be effective by virtue of this section—

(a) to discharge the interest from the mortgage, and from the operation of any order made by a court for the enforcement of the mortgage, and
(b) to extinguish any term of years created for the purposes of the mortgage,

and shall do so without the persons entitled to or interested in the mortgage or in any such order or term of years becoming parties to or executing the conveyance.

(2) Subject to subsections (3) and (4), this section applies to any mortgage of the interest so acquired (however created or arising) which—

(a) is a mortgage to secure the payment of money or the performance of any other obligation by the person from whom the interest is so acquired or any other person; and
(b) is not a mortgage which would be overreached apart from this section.

(3) This section shall not apply to any such mortgage if it has been agreed between the nominee purchaser and the reversioner or (as the case may be) any other relevant landlord that the interest in question should be acquired subject to the mortgage.

(4) In this section and Schedule 8 "mortgage" includes a charge or lien; but neither this section nor that Schedule applies to a rentcharge.

## 36 Nominee purchaser required to grant leases back to former freeholder in certain circumstances

(1) In connection with the acquisition by him of the specified premises, the nominee purchaser shall grant to the person from whom the [interest] is acquired such leases of flats or other units contained in those premises as are required to be so granted by virtue of Part II or III of Schedule 9.

(2) Any such lease shall be granted so as to take effect immediately after the acquisition by the nominee purchaser of the freehold.

(3) Where any flat or other unit demised under any such lease ("the relevant lease") is at the time of that acquisition subject to any existing lease, the relevant lease shall take effect as a lease of the freehold reversion in respect of the flat or other unit.

(4) Part IV of Schedule 9 has effect with respect to the terms of a lease granted in pursuance of Part II or III of that Schedule.

## 37 Acquisition of interests from local authorities etc

Schedule 10 to this Act (which makes provision with respect to the acquisition of interests from local authorities etc in pursuance of this Chapter) shall have effect.

## Landlord's right to compensation in relation to ineffective claims

## 37A Compensation for postponement of termination in connection with ineffective claims

(1) This section applies where a claim to exercise the right to collective enfranchisement in respect of any premises is made on or after 15th January 1999 by tenants of flats contained in the premises and the claim is not effective.

(2) A person who is a participating tenant immediately before the claim ceases to have effect shall be liable to pay compensation if—

(a) the claim was not made at least two years before the term date of the lease by virtue of which he is a qualifying tenant ("the existing lease"), and

(b) any of the conditions mentioned in subsection (3) is met.

(3) The conditions referred to above are—

(a) that the making of the claim caused a notice served under paragraph 4(1) of Schedule 10 to the Local Government and Housing Act 1989 in respect of the existing lease to cease to have effect and the date on which the claim ceases to have effect is later than four months before the termination date specified in the notice,
(b) that the making of the claim prevented the service of an effective notice under paragraph 4(1) of Schedule 10 to the Local Government and Housing Act 1989 in respect of the existing lease (but did not cause a notice served under that provision in respect of that lease to cease to have effect) and the date on which the claim ceases to have effect is a date later than six months before the term date of the existing lease, and
(c) that the existing lease has been continued under paragraph 6(1) of Schedule 3 by virtue of the claim.

(4) Compensation under subsection (2) shall become payable at the end of the appropriate period and be the right of the person who is the tenant's immediate landlord at that time.

(5) The amount which a tenant is liable to pay under subsection (2) shall be equal to the difference between—

(a) the rent for the appropriate period under the existing lease, and
(b) the rent which might reasonably be expected to be payable for that period were the property to which the existing lease relates let for a term equivalent to that period on the open market by a willing landlord on the following assumptions—
   (i) that no premium is payable in connection with the letting,
   (ii) that the letting confers no security of tenure, and
   (iii) that, except as otherwise provided by this paragraph, the letting is on the same terms as the existing lease.

(6) For the purposes of subsections (4) and (5), the appropriate period is—

(a) in a case falling within paragraph (a) of subsection (3), the period—
   (i) beginning with the termination date specified in the notice mentioned in that paragraph, and

(ii) ending with the earliest date of termination which could have been specified in a notice under paragraph 4(1) of Schedule 10 to the Local Government and Housing Act 1989 in respect of the existing lease served immediately after the date on which the claim ceases to have effect, or, if the existing lease is terminated before then, with the date of its termination;

(b) in a case falling within paragraph (b) of subsection (3), the period—

  (i) beginning with the later of six months from the date on which the claim is made and the term date of the existing lease, and
  (ii) ending six months after the date on which the claim ceases to have effect, or, if the existing lease is terminated before then, with the date of its termination; and

(c) in a case falling within paragraph (c) of subsection (3), the period for which the existing lease is continued under paragraph 6(1) of Schedule 3.

(7) In the case of a person who becomes a participating tenant by virtue of an election under section 14(3), the references in subsections (3)(a) and (b) and (6)(b)(i) to the making of the claim shall be construed as references to the making of the election.

(8) For the purposes of this section—

(a) references to a claim to exercise the right to collective enfranchisement shall be taken as references to a notice given, or purporting to be given (whether by persons who are qualifying tenants or not), under section 13,

(b) references to the date on which a claim ceases to have effect shall, in the case of a claim made by a notice which is not a valid notice under section 13, be taken as references to the date on which the notice is set aside by the court or is withdrawn or would, if valid, cease to have effect or be deemed to have been withdrawn, that date being taken, where the notice is set aside, or would, if valid, cease to have effect, in consequence of a court order, to be the date when the order becomes final, and

(c) a claim to exercise the right to collective enfranchisement is not effective if it ceases to have effect for any reason other than—

  (i) the application of section 23(4), 30(4) or 31(4),
  (ii) the entry into a binding contract for the acquisition of the freehold and other interests falling to be acquired in pursuance of the claim, or

(iii) the making of an order under section 24(4)(a) or (b) or 25(6)(a) or (b) which provides for the vesting of those interests.

## 37B Modification of section 37A where change in immediate reversion

(1) Where a tenant's liability to pay compensation under section 37A relates to a period during which there has been a change in the interest immediately expectant on the determination of his lease, that section shall have effect with the following modifications.

(2) For subsections (4) and (5) there shall be substituted—

"(4) Compensation under subsection (2) shall become payable at the end of the appropriate period and there shall be a separate right to compensation in respect of each of the interests which, during that period, have been immediately expectant on the determination of the existing lease.

(5) Compensation under subsection (2) above shall—
  (a) in the case of the interest which is immediately expectant on the determination of the existing lease at the end of the appropriate period, be the right of the person in whom that interest is vested at that time, and
  (b) in the case of an interest which ceases during the appropriate period to be immediately expectant on the determination of the existing lease, be the right of the person in whom the interest was vested immediately before it ceased to be so expectant.

(5A) The amount which the tenant is liable to pay under subsection (2) above in respect of any interest shall be equal to the difference between—

  (a) the rent under the existing lease for the part of the appropriate period during which the interest was immediately expectant on the determination of that lease, and
  (b) the rent which might reasonably be expected to be payable for that part of that period were the property to which the existing lease relates let for a term equivalent to that part of that period on the open market by a willing landlord on the following assumptions—

*Appendix*

       (i) that no premium is payable in connection with the letting,
       (ii) that the letting confers no security of tenure, and
       (iii) that, except as otherwise provided by this paragraph, the letting is on the same terms as the existing lease."

(3) In subsection (6), for "(4) and (5)" there shall be substituted "(4) to (5A)".

## Supplemental

**38 Interpretation of Chapter I**

(1) In this Chapter (unless the context otherwise requires)—

"conveyance" includes assignment, transfer and surrender, and related expressions shall be construed accordingly;
"the initial notice" means the notice given under section 13;
"introductory tenancy" has the same meaning as in Chapter I of Part V of the Housing Act 1996;
"the nominee purchaser" shall be construed in accordance with section 15;
"the participating tenants" shall be construed in accordance with section 14;
"premises with a resident landlord" shall be construed in accordance with section 10;
"public sector landlord" means any of the persons listed in section 171(2) of the Housing Act 1985;
"qualifying tenant" shall be construed in accordance with section 5;
"the relevant date" has the meaning given by section 1(8);
"relevant landlord" and "the reversioner" shall be construed in accordance with section 9;
"the right to collective enfranchisement" means the right specified in section 1(1);
"secure tenancy" has the meaning given by section 79 of the Housing Act 1985;
"the specified premises" shall be construed in accordance with section 13(12);
"the terms of acquisition" has the meaning given by section 24(8);
"unit" means—

(a) a flat;
(b) any other separate set of premises which is constructed or adapted for use for the purposes of a dwelling; or
(c) a separate set of premises let, or intended for letting, on a business lease.

(2) Any reference in this Chapter (however expressed) to the acquisition or proposed acquisition by the nominee purchaser is a reference to the acquisition or proposed acquisition by the nominee purchaser, on behalf of the participating tenants, of such freehold and other interests as fall to be so acquired under a contract entered into in pursuance of the initial notice.

(3) Any reference in this Chapter to the interest of a relevant landlord in the specified premises is a reference to the interest in those premises by virtue of which he is, in accordance with section 9(2)(b) or (2A)(b), a relevant landlord.

(4) Any reference in this Chapter to agreement in relation to all or any of the terms of acquisition is a reference to agreement subject to contract.

# The Commonhold and Leasehold Reform Act 2002
## Part 2
## Leasehold Reform

## Chapter 1
## Right to Manage

### Introductory

**71 The right to manage**
(1) This Chapter makes provision for the acquisition and exercise of rights in relation to the management of premises to which this Chapter applies by a company which, in accordance with this Chapter, may acquire and exercise those rights (referred to in this Chapter as a RTM company).

(2) The rights are to be acquired and exercised subject to and in accordance with this Chapter and are referred to in this Chapter as the right to manage.

# Qualifying rules

## 72 Premises to which Chapter applies
(1) This Chapter applies to premises if—

(a) they consist of a self-contained building or part of a building, with or without appurtenant property,
(b) they contain two or more flats held by qualifying tenants, and
(c) the total number of flats held by such tenants is not less than two-thirds of the total number of flats contained in the premises.

(2) A building is a self-contained building if it is structurally detached.

(3) A part of a building is a self-contained part of the building if—

(a) it constitutes a vertical division of the building,
(b) the structure of the building is such that it could be redeveloped independently of the rest of the building, and
(c) subsection (4) applies in relation to it.

(4) This subsection applies in relation to a part of a building if the relevant services provided for occupiers of it—

(a) are provided independently of the relevant services provided for occupiers of the rest of the building, or
(b) could be so provided without involving the carrying out of works likely to result in a significant interruption in the provision of any relevant services for occupiers of the rest of the building.

(5) Relevant services are services provided by means of pipes, cables or other fixed installations.

(6) Schedule 6 (premises excepted from this Chapter) has effect.

## 73 RTM companies
(1) This section specifies what is a RTM company.
(2) A company is a RTM company in relation to premises if—

(a) it is a private company limited by guarantee, and
(b) its memorandum of association states that its object, or one of its objects, is the acquisition and exercise of the right to manage the premises.

(3) But a company is not a RTM company if it is a commonhold association (within the meaning of Part 1).

(4) And a company is not a RTM company in relation to premises if another company is already a RTM company in relation to the premises or to any premises containing or contained in the premises.

(5) If the freehold of any premises is conveyed or transferred to a company which is a RTM company in relation to the premises, or any premises containing or contained in the premises, it ceases to be a RTM company when the conveyance or transfer is executed.

## 74 RTM companies: membership and regulations

(1) The persons who are entitled to be members of a company which is a RTM company in relation to premises are—

(a) qualifying tenants of flats contained in the premises, and
(b) from the date on which it acquires the right to manage (referred to in this Chapter as the "acquisition date"), landlords under leases of the whole or any part of the premises.

(2) The appropriate national authority shall make regulations about the content and form of the memorandum of association and articles of association of RTM companies.

(3) A RTM company may adopt provisions of the regulations for its memorandum or articles.

(4) The regulations may include provision which is to have effect for a RTM company whether or not it is adopted by the company.

(5) A provision of the memorandum or articles of a RTM company has no effect to the extent that it is inconsistent with the regulations.

(6) The regulations have effect in relation to a memorandum or articles—

(a) irrespective of the date of the memorandum or articles, but
(b) subject to any transitional provisions of the regulations.

(7) The following provisions of the Companies Act 1985 (c 6) do not apply to a RTM company—

(a) sections 2(7) and 3 (memorandum), and
(b) section 8 (articles).

## 75 Qualifying tenants

(1) This section specifies whether there is a qualifying tenant of a flat for the purposes of this Chapter and, if so, who it is.

(2) Subject as follows, a person is the qualifying tenant of a flat if he is tenant of the flat under a long lease.

(3) Subsection (2) does not apply where the lease is a tenancy to which Part 2 of the Landlord and Tenant Act 1954 (c 56) (business tenancies) applies.

(4) Subsection (2) does not apply where—

(a) the lease was granted by sub-demise out of a superior lease other than a long lease,
(b) the grant was made in breach of the terms of the superior lease, and
(c) there has been no waiver of the breach by the superior landlord.

(5) No flat has more than one qualifying tenant at any one time; and subsections (6) and (7) apply accordingly.

(6) Where a flat is being let under two or more long leases, a tenant under any of those leases which is superior to that held by another is not the qualifying tenant of the flat.

(7) Where a flat is being let to joint tenants under a long lease, the joint tenants shall (subject to subsection (6)) be regarded as jointly being the qualifying tenant of the flat.

## 76 Long leases

(1) This section and section 77 specify what is a long lease for the purposes of this Chapter.

(2) Subject to section 77, a lease is a long lease if—

(a) it is granted for a term of years certain exceeding 21 years, whether or not it is (or may become) terminable before the end of that term by notice given by or to the tenant, by re-entry or forfeiture or otherwise,
(b) it is for a term fixed by law under a grant with a covenant or obligation for perpetual renewal (but is not a lease by sub-demise from one which is not a long lease),
(c) it takes effect under section 149(6) of the Law of Property Act 1925 (c 20) (leases terminable after a death or marriage [or the formation of a civil partnership]),
(d) it was granted in pursuance of the right to buy conferred by Part 5 of the Housing Act 1985 (c 68) or in pursuance of the right to acquire on rent to mortgage terms conferred by that Part of that Act,
(e) it is a shared ownership lease, whether granted in pursuance of that Part of that Act or otherwise, where the tenant's total share is 100 per cent, or
(f) it was granted in pursuance of that Part of that Act as it has effect by virtue of section 17 of the Housing Act 1996 (c 52) (the right to acquire).

(3) "Shared ownership lease" means a lease—

(a) granted on payment of a premium calculated by reference to a percentage of the value of the demised premises or the cost of providing them, or
(b) under which the tenant (or his personal representatives) will or may be entitled to a sum calculated by reference, directly or indirectly, to the value of those premises.

(4) "Total share", in relation to the interest of a tenant under a shared ownership lease, means his initial share plus any additional share or shares in the demised premises which he has acquired.

**77 Long leases: further provisions**
(1) A lease terminable by notice after [a death, a marriage or the formation of a civil partnership] is not a long lease if—

(a) the notice is capable of being given at any time after the death or marriage of[, or the formation of a civil partnership by,] the tenant,

(b) the length of the notice is not more than three months, and
(c) the terms of the lease preclude both its assignment otherwise than by virtue of section 92 of the Housing Act 1985 (assignments by way of exchange) and the sub-letting of the whole of the demised premises.

(2) Where the tenant of any property under a long lease, on the coming to an end of the lease, becomes or has become tenant of the property or part of it under any subsequent tenancy (whether by express grant or by implication of law), that tenancy is a long lease irrespective of its terms.

(3) A lease—
(a) granted for a term of years certain not exceeding 21 years, but with a covenant or obligation for renewal without payment of a premium (but not for perpetual renewal), and
(b) renewed on one or more occasions so as to bring to more than 21 years the total of the terms granted (including any interval between the end of a lease and the grant of a renewal),

is to be treated as if the term originally granted had been one exceeding 21 years.

(4) Where a long lease—
(a) is or was continued for any period under Part 1 of the Landlord and Tenant Act 1954 (c 56) or under Schedule 10 to the Local Government and Housing Act 1989 (c 42), or
(b) was continued for any period under the Leasehold Property (Temporary Provisions) Act 1951 (c 38),

it remains a long lease during that period.

(5) Where in the case of a flat there are at any time two or more separate leases, with the same landlord and the same tenant, and—
(a) the property comprised in one of those leases consists of either the flat or a part of it (in either case with or without appurtenant property), and
(b) the property comprised in every other lease consists of either a part of the flat (with or without appurtenant property) or appurtenant property only,

there shall be taken to be a single long lease of the property comprised in such of those leases as are long leases.

## Claim to acquire right

### 78 Notice inviting participation

(1) Before making a claim to acquire the right to manage any premises, a RTM company must give notice to each person who at the time when the notice is given—

(a) is the qualifying tenant of a flat contained in the premises, but
(b) neither is nor has agreed to become a member of the RTM company.

(2) A notice given under this section (referred to in this Chapter as a "notice of invitation to participate") must—

(a) state that the RTM company intends to acquire the right to manage the premises,
(b) state the names of the members of the RTM company,
(c) invite the recipients of the notice to become members of the company, and
(d) contain such other particulars (if any) as may be required to be contained in notices of invitation to participate by regulations made by the appropriate national authority.

(3) A notice of invitation to participate must also comply with such requirements (if any) about the form of notices of invitation to participate as may be prescribed by regulations so made.

(4) A notice of invitation to participate must either—

(a) be accompanied by a copy of the memorandum of association and articles of association of the RTM company, or
(b) include a statement about inspection and copying of the memorandum of association and articles of association of the RTM company.

(5) A statement under subsection (4)(b) must—

(a) specify a place (in England or Wales) at which the memorandum of association and articles of association may be inspected,

*Appendix*

(b) specify as the times at which they may be inspected periods of at least two hours on each of at least three days (including a Saturday or Sunday or both) within the seven days beginning with the day following that on which the notice is given,
(c) specify a place (in England or Wales) at which, at any time within those seven days, a copy of the memorandum of association and articles of association may be ordered, and
(d) specify a fee for the provision of an ordered copy, not exceeding the reasonable cost of providing it.

(6) Where a notice given to a person includes a statement under subsection (4)(b), the notice is to be treated as not having been given to him if he is not allowed to undertake an inspection, or is not provided with a copy, in accordance with the statement.

(7) A notice of invitation to participate is not invalidated by any inaccuracy in any of the particulars required by or by virtue of this section.

**79 Notice of claim to acquire right**
(1) A claim to acquire the right to manage any premises is made by giving notice of the claim (referred to in this Chapter as a "claim notice"); and in this Chapter the "relevant date", in relation to any claim to acquire the right to manage, means the date on which notice of the claim is given.

(2) The claim notice may not be given unless each person required to be given a notice of invitation to participate has been given such a notice at least 14 days before.

(3) The claim notice must be given by a RTM company which complies with subsection (4) or (5).

(4) If on the relevant date there are only two qualifying tenants of flats contained in the premises, both must be members of the RTM company.

(5) In any other case, the membership of the RTM company must on the relevant date include a number of qualifying tenants of flats contained in the premises which is not less than one-half of the total number of flats so contained.

(6) The claim notice must be given to each person who on the relevant date is—

(a) landlord under a lease of the whole or any part of the premises,
(b) party to such a lease otherwise than as landlord or tenant, or
(c) a manager appointed under Part 2 of the Landlord and Tenant Act 1987 (c 31) (referred to in this Part as "the 1987 Act") to act in relation to the premises, or any premises containing or contained in the premises.

(7) Subsection (6) does not require the claim notice to be given to a person who cannot be found or whose identity cannot be ascertained; but if this subsection means that the claim notice is not required to be given to anyone at all, section 85 applies.

(8) A copy of the claim notice must be given to each person who on the relevant date is the qualifying tenant of a flat contained in the premises.

(9) Where a manager has been appointed under Part 2 of the 1987 Act to act in relation to the premises, or any premises containing or contained in the premises, a copy of the claim notice must also be given to the leasehold valuation tribunal or court by which he was appointed.

**80 Contents of claim notice**

(1) The claim notice must comply with the following requirements.

(2) It must specify the premises and contain a statement of the grounds on which it is claimed that they are premises to which this Chapter applies.

(3) It must state the full name of each person who is both—

(a) the qualifying tenant of a flat contained in the premises, and
(b) a member of the RTM company,

and the address of his flat.

(4) And it must contain, in relation to each such person, such particulars of his lease as are sufficient to identify it, including—

(a) the date on which it was entered into,
(b) the term for which it was granted, and
(c) the date of the commencement of the term.

(5)  It must state the name and registered office of the RTM company.

(6)  It must specify a date, not earlier than one month after the relevant date, by which each person who was given the notice under section 79(6) may respond to it by giving a counter-notice under section 84.

(7)  It must specify a date, at least three months after that specified under subsection (6), on which the RTM company intends to acquire the right to manage the premises.

(8)  It must also contain such other particulars (if any) as may be required to be contained in claim notices by regulations made by the appropriate national authority.

(9)  And it must comply with such requirements (if any) about the form of claim notices as may be prescribed by regulations so made.

**81 Claim notice: supplementary**

(1)  A claim notice is not invalidated by any inaccuracy in any of the particulars required by or by virtue of section 80.

(2)  Where any of the members of the RTM company whose names are stated in the claim notice was not the qualifying tenant of a flat contained in the premises on the relevant date, the claim notice is not invalidated on that account, so long as a sufficient number of qualifying tenants of flats contained in the premises were members of the company on that date; and for this purpose a "sufficient number" is a number (greater than one) which is not less than one-half of the total number of flats contained in the premises on that date.

(3)  Where any premises have been specified in a claim notice, no subsequent claim notice which specifies—

(a) the premises, or
(b) any premises containing or contained in the premises,

may be given so long as the earlier claim notice continues in force.

(4) Where a claim notice is given by a RTM company it continues in force from the relevant date until the right to manage is acquired by the company unless it has previously—

(a) been withdrawn or deemed to be withdrawn by virtue of any provision of this Chapter, or
(b) ceased to have effect by reason of any other provision of this Chapter.

## 82 Right to obtain information

(1) A company which is a RTM company in relation to any premises may give to any person a notice requiring him to provide the company with any information—

(a) which is in his possession or control, and
(b) which the company reasonably requires for ascertaining the particulars required by or by virtue of section 80 to be included in a claim notice for claiming to acquire the right to manage the premises.

(2) Where the information is recorded in a document in the person's possession or control, the RTM company may give him a notice requiring him—

(a) to permit any person authorised to act on behalf of the company at any reasonable time to inspect the document (or, if the information is recorded in the document in a form in which it is not readily intelligible, to give any such person access to it in a readily intelligible form), and
(b) to supply the company with a copy of the document containing the information in a readily intelligible form on payment of a reasonable fee.

(3) A person to whom a notice is given must comply with it within the period of 28 days beginning with the day on which it is given.

## 83 Right of access

(1) Where a RTM company has given a claim notice in relation to any premises, each of the persons specified in subsection (2) has a right of access to any part of the premises if that is reasonable in connection with any matter arising out of the claim to acquire the right to manage.

(2) The persons referred to in subsection (1) are—

(a) any person authorised to act on behalf of the RTM company,
(b) any person who is landlord under a lease of the whole or any part of the premises and any person authorised to act on behalf of any such person,
(c) any person who is party to such a lease otherwise than as landlord or tenant and any person authorised to act on behalf of any such person, and
(d) any manager appointed under Part 2 of the 1987 Act to act in relation to the premises, or any premises containing or contained in the premises, and any person authorised to act on behalf of any such manager.

(3) The right conferred by this section is exercisable, at any reasonable time, on giving not less than ten days' notice—

(a) to the occupier of any premises to which access is sought, or
(b) if those premises are unoccupied, to the person entitled to occupy them.

**84 Counter-notices**

(1) A person who is given a claim notice by a RTM company under section 79(6) may give a notice (referred to in this Chapter as a "counter-notice") to the company no later than the date specified in the claim notice under section 80(6).

(2) A counter-notice is a notice containing a statement either—

(a) admitting that the RTM company was on the relevant date entitled to acquire the right to manage the premises specified in the claim notice, or
(b) alleging that, by reason of a specified provision of this Chapter, the RTM company was on that date not so entitled,

and containing such other particulars (if any) as may be required to be contained in counter-notices, and complying with such requirements (if any) about the form of counter-notices, as may be prescribed by regulations made by the appropriate national authority.

(3) Where the RTM company has been given one or more counter-notices containing a statement such as is mentioned in subsection

(2)(b), the company may apply to a leasehold valuation tribunal for a determination that it was on the relevant date entitled to acquire the right to manage the premises.

(4) An application under subsection (3) must be made not later than the end of the period of two months beginning with the day on which the counter-notice (or, where more than one, the last of the counter-notices) was given.

(5) Where the RTM company has been given one or more counter-notices containing a statement such as is mentioned in subsection (2)(b), the RTM company does not acquire the right to manage the premises unless—

(a) on an application under subsection (3) it is finally determined that the company was on the relevant date entitled to acquire the right to manage the premises, or
(b) the person by whom the counter-notice was given agrees, or the persons by whom the counter-notices were given agree, in writing that the company was so entitled.

(6) If on an application under subsection (3) it is finally determined that the company was not on the relevant date entitled to acquire the right to manage the premises, the claim notice ceases to have effect.

(7) A determination on an application under subsection (3) becomes final—

(a) if not appealed against, at the end of the period for bringing an appeal, or
(b) if appealed against, at the time when the appeal (or any further appeal) is disposed of.

(8) An appeal is disposed of—

(a) if it is determined and the period for bringing any further appeal has ended, or
(b) if it is abandoned or otherwise ceases to have effect.

## 85 Landlords etc not traceable

(1) This section applies where a RTM company wishing to acquire the right to manage premises—

(a) complies with subsection (4) or (5) of section 79, and
(b) would not have been precluded from giving a valid notice under that section with respect to the premises,

but cannot find, or ascertain the identity of, any of the persons to whom the claim notice would be required to be given by subsection (6) of that section.

(2) The RTM company may apply to a leasehold valuation tribunal for an order that the company is to acquire the right to manage the premises.

(3) Such an order may be made only if the company has given notice of the application to each person who is the qualifying tenant of a flat contained in the premises.

(4) Before an order is made the company may be required to take such further steps by way of advertisement or otherwise as is determined proper for the purpose of tracing the persons who are—

(a) landlords under leases of the whole or any part of the premises, or
(b) parties to such leases otherwise than as landlord or tenant.

(5) If any of those persons is traced—

(a) after an application for an order is made, but
(b) before the making of an order,

no further proceedings shall be taken with a view to the making of an order.

(6) Where that happens—

(a) the rights and obligations of all persons concerned shall be determined as if the company had, at the date of the application, duly given notice under section 79 of its claim to acquire the right to manage the premises, and
(b) the leasehold valuation tribunal may give such directions as it thinks fit as to the steps to be taken for giving effect to their rights and obligations, including directions modifying or dispensing with any of the requirements imposed by or by virtue of this Chapter.

(7) An application for an order may be withdrawn at any time before an order is made and, after it is withdrawn, subsection (6)(a) does not apply.

(8) But where any step is taken for the purpose of giving effect to subsection (6)(a) in the case of any application, the application shall not afterwards be withdrawn except—

(a) with the consent of the person or persons traced, or
(b) by permission of the leasehold valuation tribunal.

(9) And permission shall be given only where it appears just that it should be given by reason of matters coming to the knowledge of the RTM company in consequence of the tracing of the person or persons traced.

## 86 Withdrawal of claim notice

(1) A RTM company which has given a claim notice in relation to any premises may, at any time before it acquires the right to manage the premises, withdraw the claim notice by giving a notice to that effect (referred to in this Chapter as a "notice of withdrawal").

(2) A notice of withdrawal must be given to each person who is—

(a) landlord under a lease of the whole or any part of the premises,
(b) party to such a lease otherwise than as landlord or tenant,
(c) a manager appointed under Part 2 of the 1987 Act to act in relation to the premises, or any premises containing or contained in the premises, or
(d) the qualifying tenant of a flat contained in the premises.

## 87 Deemed withdrawal

(1) If a RTM company has been given one or more counter-notices containing a statement such as is mentioned in subsection (2)(b) of section 84 but either—

(a) no application for a determination under subsection (3) of that section is made within the period specified in subsection (4) of that section, or
(b) such an application is so made but is subsequently withdrawn,

the claim notice is deemed to be withdrawn.

(2) The withdrawal shall be taken to occur—

(a) if paragraph (a) of subsection (1) applies, at the end of the period specified in that paragraph, and
(b) if paragraph (b) of that subsection applies, on the date of the withdrawal of the application.

(3) Subsection (1) does not apply if the person by whom the counter-notice was given has, or the persons by whom the counter-notices were given have, (before the time when the withdrawal would be taken to occur) agreed in writing that the RTM company was on the relevant date entitled to acquire the right to manage the premises.

(4) The claim notice is deemed to be withdrawn if—

(a) a winding-up order ... is made, or a resolution for voluntary winding-up is passed, with respect to the RTM company, [or the RTM company enters administration,]
(b) a receiver or a manager of the RTM company's undertaking is duly appointed, or possession is taken, by or on behalf of the holders of any debentures secured by a floating charge, of any property of the RTM company comprised in or subject to the charge,
(c) a voluntary arrangement proposed in the case of the RTM company for the purposes of Part 1 of the Insolvency Act 1986 (c 45) is approved under that Part of that Act, or
(d) the RTM company's name is struck off the register under section 652 or 652A of the Companies Act 1985 (c 6).

## 88 Costs: general

(1) A RTM company is liable for reasonable costs incurred by a person who is—

(a) landlord under a lease of the whole or any part of any premises,
(b) party to such a lease otherwise than as landlord or tenant, or
(c) a manager appointed under Part 2 of the 1987 Act to act in relation to the premises, or any premises containing or contained in the premises,

in consequence of a claim notice given by the company in relation to the premises.

(2) Any costs incurred by such a person in respect of professional services rendered to him by another are to be regarded as reasonable only if and to the extent that costs in respect of such services might reasonably be expected to have been incurred by him if the circumstances had been such that he was personally liable for all such costs.

(3) A RTM company is liable for any costs which such a person incurs as party to any proceedings under this Chapter before a leasehold valuation tribunal only if the tribunal dismisses an application by the company for a determination that it is entitled to acquire the right to manage the premises.

(4) Any question arising in relation to the amount of any costs payable by a RTM company shall, in default of agreement, be determined by a leasehold valuation tribunal.

## 89 Costs where claim ceases

(1) This section applies where a claim notice given by a RTM company—

(a) is at any time withdrawn or deemed to be withdrawn by virtue of any provision of this Chapter, or
(b) at any time ceases to have effect by reason of any other provision of this Chapter.

(2) The liability of the RTM company under section 88 for costs incurred by any person is a liability for costs incurred by him down to that time.

(3) Each person who is or has been a member of the RTM company is also liable for those costs (jointly and severally with the RTM company and each other person who is so liable).

(4) But subsection (3) does not make a person liable if—

(a) the lease by virtue of which he was a qualifying tenant has been assigned to another person, and
(b) that other person has become a member of the RTM company.

(5) The reference in subsection (4) to an assignment includes—

(a) an assent by personal representatives, and
(b) assignment by operation of law where the assignment is to a trustee in bankruptcy or to a mortgagee under section 89(2) of the Law of Property Act 1925 (c 20) (foreclosure of leasehold mortgage).

## Acquisition of right

### 90 The acquisition date

(1) This section makes provision about the date which is the acquisition date where a RTM company acquires the right to manage any premises.

(2) Where there is no dispute about entitlement, the acquisition date is the date specified in the claim notice under section 80(7).

(3) For the purposes of this Chapter there is no dispute about entitlement if—
(a) no counter-notice is given under section 84, or
(b) the counter-notice given under that section, or (where more than one is so given) each of them, contains a statement such as is mentioned in subsection (2)(a) of that section.

(4) Where the right to manage the premises is acquired by the company by virtue of a determination under section 84(5)(a), the acquisition date is the date three months after the determination becomes final.

(5) Where the right to manage the premises is acquired by the company by virtue of subsection (5)(b) of section 84, the acquisition date is the date three months after the day on which the person (or the last person) by whom a counter-notice containing a statement such as is mentioned in subsection (2)(b) of that section was given agrees in writing that the company was on the relevant date entitled to acquire the right to manage the premises.

(6) Where an order is made under section 85, the acquisition date is (subject to any appeal) the date specified in the order.

## 91 Notices relating to management contracts
(1) Section 92 applies where—

(a) the right to manage premises is to be acquired by a RTM company (otherwise than by virtue of an order under section 85), and
(b) there are one or more existing management contracts relating to the premises.

(2) A management contract is a contract between—

(a) an existing manager of the premises (referred to in this Chapter as the "manager party"), and
(b) another person (so referred to as the "contractor party"),

under which the contractor party agrees to provide services, or do any other thing, in connection with any matter relating to a function which will be a function of the RTM company once it acquires the right to manage.

(3) And in this Chapter "existing management contract" means a management contract which—

(a) is subsisting immediately before the determination date, or
(b) is entered into during the period beginning with the determination date and ending with the acquisition date.

(4) An existing manager of the premises is any person who is—

(a) landlord under a lease relating to the whole or any part of the premises,
(b) party to such a lease otherwise than as landlord or tenant, or
(c) a manager appointed under Part 2 of the 1987 Act to act in relation to the premises, or any premises containing or contained in the premises.

(5) In this Chapter "determination date" means—

(a) where there is no dispute about entitlement, the date specified in the claim notice under section 80(6),
(b) where the right to manage the premises is acquired by the company by virtue of a determination under section 84(5)(a), the date when the determination becomes final, and
(c) where the right to manage the premises is acquired by the company by virtue of subsection (5)(b) of section 84, the day on which the

person (or the last person) by whom a counter-notice containing a statement such as is mentioned in subsection (2)(b) of that section was given agrees in writing that the company was on the relevant date entitled to acquire the right to manage the premises.

## 92 Duties to give notice of contracts

(1) The person who is the manager party in relation to an existing management contract must give a notice in relation to the contract—

(a) to the person who is the contractor party in relation to the contract (a "contractor notice"), and
(b) to the RTM company (a "contract notice").

(2) A contractor notice and a contract notice must be given—

(a) in the case of a contract subsisting immediately before the determination date, on that date or as soon after that date as is reasonably practicable, and
(b) in the case of a contract entered into during the period beginning with the determination date and ending with the acquisition date, on the date on which it is entered into or as soon after that date as is reasonably practicable.

(3) A contractor notice must—

(a) give details sufficient to identify the contract in relation to which it is given,
(b) state that the right to manage the premises is to be acquired by a RTM company,
(c) state the name and registered office of the RTM company,
(d) specify the acquisition date, and
(e) contain such other particulars (if any) as may be required to be contained in contractor notices by regulations made by the appropriate national authority,

and must also comply with such requirements (if any) about the form of contractor notices as may be prescribed by regulations so made.

(4) Where a person who receives a contractor notice (including one who receives a copy by virtue of this subsection) is party to an existing management sub-contract with another person (the "sub-contractor party"), the person who received the notice must—

(a) send a copy of the contractor notice to the sub-contractor party, and
(b) give to the RTM company a contract notice in relation to the existing management sub-contract.

(5) An existing management sub-contract is a contract under which the sub-contractor party agrees to provide services, or do any other thing, in connection with any matter relating to a function which will be a function of the RTM company once it acquires the right to manage and which—
(a) is subsisting immediately before the determination date, or
(b) is entered into during the period beginning with the determination date and ending with the acquisition date.

(6) Subsection (4) must be complied with—
(a) in the case of a contract entered into before the contractor notice is received, on the date on which it is received or as soon after that date as is reasonably practicable, and
(b) in the case of a contract entered into after the contractor notice is received, on the date on which it is entered into or as soon after that date as is reasonably practicable.

(7) A contract notice must—
(a) give particulars of the contract in relation to which it is given and of the person who is the contractor party, or sub-contractor party, in relation to that contract, and
(b) contain such other particulars (if any) as may be required to be contained in contract notices by regulations made by the appropriate national authority,

and must also comply with such requirements (if any) about the form of contract notices as may be prescribed by such regulations so made.

## 93 Duty to provide information
(1) Where the right to manage premises is to be acquired by a RTM company, the company may give notice to a person who is—
(a) landlord under a lease of the whole or any part of the premises,
(b) party to such a lease otherwise than as landlord or tenant, or

(c) a manager appointed under Part 2 of the 1987 Act to act in relation to the premises, or any premises containing or contained in the premises,

requiring him to provide the company with any information which is in his possession or control and which the company reasonably requires in connection with the exercise of the right to manage.

(2) Where the information is recorded in a document in his possession or control the notice may require him—

(a) to permit any person authorised to act on behalf of the company at any reasonable time to inspect the document (or, if the information is recorded in the document in a form in which it is not readily intelligible, to give any such person access to it in a readily intelligible form), and
(b) to supply the company with a copy of the document containing the information in a readily intelligible form.

(3) A notice may not require a person to do anything under this section before the acquisition date.

(4) But, subject to that, a person who is required by a notice to do anything under this section must do it within the period of 28 days beginning with the day on which the notice is given.

## 94 Duty to pay accrued uncommitted service charges

(1) Where the right to manage premises is to be acquired by a RTM company, a person who is—

(a) landlord under a lease of the whole or any part of the premises,
(b) party to such a lease otherwise than as landlord or tenant, or
(c) a manager appointed under Part 2 of the 1987 Act to act in relation to the premises, or any premises containing or contained in the premises,

must make to the company a payment equal to the amount of any accrued uncommitted service charges held by him on the acquisition date.

(2) The amount of any accrued uncommitted service charges is the aggregate of—

(a) any sums which have been paid to the person by way of service charges in respect of the premises, and
(b) any investments which represent such sums (and any income which has accrued on them),

less so much (if any) of that amount as is required to meet the costs incurred before the acquisition date in connection with the matters for which the service charges were payable.

(3) He or the RTM company may make an application to a leasehold valuation tribunal to determine the amount of any payment which falls to be made under this section.

(4) The duty imposed by this section must be complied with on the acquisition date or as soon after that date as is reasonably practicable.

## Exercising right

**95 Introductory**
Sections 96 to 103 apply where the right to manage premises has been acquired by a RTM company (and has not ceased to be exercisable by it).

**96 Management functions under leases**
(1) This section and section 97 apply in relation to management functions relating to the whole or any part of the premises.

(2) Management functions which a person who is landlord under a lease of the whole or any part of the premises has under the lease are instead functions of the RTM company.

(3) And where a person is party to a lease of the whole or any part of the premises otherwise than as landlord or tenant, management functions of his under the lease are also instead functions of the RTM company.

(4) Accordingly, any provisions of the lease making provision about the relationship of—

(a) a person who is landlord under the lease, and

(b) a person who is party to the lease otherwise than as landlord or tenant,

in relation to such functions do not have effect.

(5) "Management functions" are functions with respect to services, repairs, maintenance, improvements, insurance and management.

(6) But this section does not apply in relation to—

(a) functions with respect to a matter concerning only a part of the premises consisting of a flat or other unit not held under a lease by a qualifying tenant, or
(b) functions relating to re-entry or forfeiture.

(7) An order amending subsection (5) or (6) may be made by the appropriate national authority.

## 97 Management functions: supplementary

(1) Any obligation owed by the RTM company by virtue of section 96 to a tenant under a lease of the whole or any part of the premises is also owed to each person who is landlord under the lease.

(2) A person who is—

(a) landlord under a lease of the whole or any part of the premises,
(b) party to such a lease otherwise than as landlord or tenant, or
(c) a manager appointed under Part 2 of the 1987 Act to act in relation to the premises, or any premises containing or contained in the premises,

is not entitled to do anything which the RTM company is required or empowered to do under the lease by virtue of section 96, except in accordance with an agreement made by him and the RTM company.

(3) But subsection (2) does not prevent any person from insuring the whole or any part of the premises at his own expense.

(4) So far as any function of a tenant under a lease of the whole or any part of the premises—

(a) relates to the exercise of any function under the lease which is a function of the RTM company by virtue of section 96, and

(b) is exercisable in relation to a person who is landlord under the lease or party to the lease otherwise than as landlord or tenant,

it is instead exercisable in relation to the RTM company.

(5) But subsection (4) does not require or permit the payment to the RTM company of so much of any service charges payable by a tenant under a lease of the whole or any part of the premises as is required to meet costs incurred before the right to manage was acquired by the RTM company in connection with matters for which the service charges are payable.

**98 Functions relating to approvals**

(1) This section and section 99 apply in relation to the grant of approvals under long leases of the whole or any part of the premises; but nothing in this section or section 99 applies in relation to an approval concerning only a part of the premises consisting of a flat or other unit not held under a lease by a qualifying tenant.

(2) Where a person who is—

(a) landlord under a long lease of the whole or any part of the premises, or
(b) party to such a lease otherwise than as landlord or tenant,

has functions in relation to the grant of approvals to a tenant under the lease, the functions are instead functions of the RTM company.

(3) Accordingly, any provisions of the lease making provision about the relationship of—

(a) a person who is landlord under the lease, and
(b) a person who is party to the lease otherwise than as landlord or tenant,

in relation to such functions do not have effect.

(4) The RTM company must not grant an approval by virtue of subsection (2) without having given—

(a) in the case of an approval relating to assignment, underletting, charging, parting with possession, the making of structural alterations or improvements or alterations of use, 30 days' notice, or

(b) in any other case, 14 days' notice,

to the person who is, or each of the persons who are, landlord under the lease.

(5) Regulations increasing the period of notice to be given under subsection (4)(b) in the case of any description of approval may be made by the appropriate national authority.

(6) So far as any function of a tenant under a long lease of the whole or any part of the premises—

(a) relates to the exercise of any function which is a function of the RTM company by virtue of this section, and
(b) is exercisable in relation to a person who is landlord under the lease or party to the lease otherwise than as landlord or tenant,

it is instead exercisable in relation to the RTM company.

(7) In this Chapter "approval" includes consent or licence and "approving" is to be construed accordingly; and an approval required to be obtained by virtue of a restriction entered on the register of title kept by the Chief Land Registrar is, so far as relating to a long lease of the whole or any part of any premises, to be treated for the purposes of this Chapter as an approval under the lease.

**99 Approvals: supplementary**
(1) If a person to whom notice is given under section 98(4) objects to the grant of the approval before the time when the RTM company would first be entitled to grant it, the RTM company may grant it only—

(a) in accordance with the written agreement of the person who objected, or
(b) in accordance with a determination of (or on an appeal from) a leasehold valuation tribunal.

(2) An objection to the grant of the approval may not be made by a person unless he could withhold the approval if the function of granting it were exercisable by him (and not by the RTM company).

(3) And a person may not make an objection operating only if a condition or requirement is not satisfied unless he could grant the

approval subject to the condition or requirement being satisfied if the function of granting it were so exercisable.

(4) An objection to the grant of the approval is made by giving notice of the objection (and of any condition or requirement which must be satisfied if it is not to operate) to—

(a) the RTM company, and
(b) the tenant,

and, if the approval is to a tenant approving an act of a sub-tenant, to the sub-tenant.

(5) An application to a leasehold valuation tribunal for a determination under subsection (1)(b) may be made by—

(a) the RTM company,
(b) the tenant,
(c) if the approval is to a tenant approving an act of a sub-tenant, the sub-tenant, or
(d) any person who is landlord under the lease.


**100 Enforcement of tenant covenants**
(1) This section applies in relation to the enforcement of untransferred tenant covenants of a lease of the whole or any part of the premises.

(2) Untransferred tenant covenants are enforceable by the RTM company, as well as by any other person by whom they are enforceable apart from this section, in the same manner as they are enforceable by any other such person.

(3) But the RTM company may not exercise any function of re-entry or forfeiture.

(4) In this Chapter "tenant covenant", in relation to a lease, means a covenant falling to be complied with by a tenant under the lease; and a tenant covenant is untransferred if, apart from this section, it would not be enforceable by the RTM company.

(5) Any power under a lease of a person who is—

(a) landlord under the lease, or

(b) party to the lease otherwise than as landlord or tenant,

to enter any part of the premises to determine whether a tenant is complying with any untransferred tenant covenant is exercisable by the RTM company (as well as by the landlord or party).

**101 Tenant covenants: monitoring and reporting**
(1) This section applies in relation to failures to comply with tenant covenants of leases of the whole or any part of the premises.

(2) The RTM company must—

(a) keep under review whether tenant covenants of leases of the whole or any part of the premises are being complied with, and
(b) report to any person who is landlord under such a lease any failure to comply with any tenant covenant of the lease.

(3) The report must be made before the end of the period of three months beginning with the day on which the failure to comply comes to the attention of the RTM company.

(4) But the RTM company need not report to a landlord a failure to comply with a tenant covenant if—

(a) the failure has been remedied,
(b) reasonable compensation has been paid in respect of the failure, or
(c) the landlord has notified the RTM company that it need not report to him failures of the description of the failure concerned.

**102 Statutory functions**
(1) Schedule 7 (provision for the operation of certain enactments with modifications) has effect.

(2) Other enactments relating to leases (including enactments contained in this Act or any Act passed after this Act) have effect with any such modifications as are prescribed by regulations made by the appropriate national authority.

## 103 Landlord contributions to service charges

(1) This section applies where—

(a) the premises contain at least one flat or other unit not subject to a lease held by a qualifying tenant (an "excluded unit"),
(b) the service charges payable under leases of flats contained in the premises which are so subject fall to be calculated as a proportion of the relevant costs, and
(c) the proportions of the relevant costs so payable, when aggregated, amount to less than the whole of the relevant costs.

(2) Where the premises contain only one excluded unit, the person who is the appropriate person in relation to the excluded unit must pay to the RTM company the difference between—

(a) the relevant costs, and
(b) the aggregate amount payable in respect of the relevant costs under leases of flats contained in the premises which are held by qualifying tenants.

(3) Where the premises contain more than one excluded unit, each person who is the appropriate person in relation to an excluded unit must pay to the RTM company the appropriate proportion of that difference.

(4) And the appropriate proportion in the case of each such person is the proportion of the internal floor area of all of the excluded units which is internal floor area of the excluded unit in relation to which he is the appropriate person.

(5) The appropriate person in relation to an excluded unit—

(a) if it is subject to a lease, is the landlord under the lease,
(b) if it is subject to more than one lease, is the immediate landlord under whichever of the leases is inferior to all the others, and
(c) if it is not subject to any lease, is the freeholder.

## Supplementary

**104** ...

**105 Cessation of management**

(1) This section makes provision about the circumstances in which, after a RTM company has acquired the right to manage any premises, that right ceases to be exercisable by it.

(2) Provision may be made by an agreement made between—

(a) the RTM company, and
(b) each person who is landlord under a lease of the whole or any part of the premises,

for the right to manage the premises to cease to be exercisable by the RTM company.

(3) The right to manage the premises ceases to be exercisable by the RTM company if—

(a) a winding-up order ... is made, or a resolution for voluntary winding-up is passed, with respect to the RTM company, [or the RTM company enters administration,]
(b) a receiver or a manager of the RTM company's undertaking is duly appointed, or possession is taken, by or on behalf of the holders of any debentures secured by a floating charge, of any property of the RTM company comprised in or subject to the charge,
(c) a voluntary arrangement proposed in the case of the RTM company for the purposes of Part 1 of the Insolvency Act 1986 (c 45) is approved under that Part of that Act, or
(d) the RTM company's name is struck off the register under section 652 or 652A of the Companies Act 1985 (c 6).

(4) The right to manage the premises ceases to be exercisable by the RTM company if a manager appointed under Part 2 of the 1987 Act to act in relation to the premises, or any premises containing or contained in the premises, begins so to act or an order under that Part of that Act that the right to manage the premises is to cease to be exercisable by the RTM company takes effect.

(5) The right to manage the premises ceases to be exercisable by the RTM company if it ceases to be a RTM company in relation to the premises.

## 106 Agreements excluding or modifying right

Any agreement relating to a lease (whether contained in the instrument creating the lease or not and whether made before the creation of the lease or not) is void in so far as it—

(a) purports to exclude or modify the right of any person to be, or do any thing as, a member of a RTM company,
(b) provides for the termination or surrender of the lease if the tenant becomes, or does any thing as, a member of a RTM company or if a RTM company does any thing, or
(c) provides for the imposition of any penalty or disability if the tenant becomes, or does any thing as, a member of a RTM company or if a RTM company does any thing.

## 107 Enforcement of obligations

(1) A county court may, on the application of any person interested, make an order requiring a person who has failed to comply with a requirement imposed on him by, under or by virtue of any provision of this Chapter to make good the default within such time as is specified in the order.

(2) An application shall not be made under subsection (1) unless—

(a) a notice has been previously given to the person in question requiring him to make good the default, and
(b) more than 14 days have elapsed since the date of the giving of that notice without his having done so.

## 108 Application to Crown

(1) This Chapter applies in relation to premises in which there is a Crown interest.

(2) There is a Crown interest in premises if there is in the premises an interest or estate—

(a) which is comprised in the Crown Estate,
(b) which belongs to Her Majesty in right of the Duchy of Lancaster,
(c) which belongs to the Duchy of Cornwall, or
(d) which belongs to a government department or is held on behalf of Her Majesty for the purposes of a government department.

(3) Any sum payable under this Chapter to a RTM company by the Chancellor of the Duchy of Lancaster may be raised and paid under section 25 of the Duchy of Lancaster Act 1817 (c 97) as an expense incurred in improvement of land belonging to Her Majesty in right of the Duchy.

(4) Any sum payable under this Chapter to a RTM company by the Duke of Cornwall (or any other possessor for the time being of the Duchy of Cornwall) may be raised and paid under section 8 of the Duchy of Cornwall Management Act 1863 (c 49) as an expense incurred in permanently improving the possessions of the Duchy.

**109 Powers of trustees in relation to right**
(1) Where trustees are the qualifying tenant of a flat contained in any premises, their powers under the instrument regulating the trusts include power to be a member of a RTM company for the purpose of the acquisition and exercise of the right to manage the premises.

(2) But subsection (1) does not apply where the instrument regulating the trusts contains an explicit direction to the contrary.

(3) The power conferred by subsection (1) is exercisable with the same consent or on the same direction (if any) as may be required for the exercise of the trustees' powers (or ordinary powers) of investment.

(4) The purposes—

(a) authorised for the application of capital money by section 73 of the Settled Land Act 1925 (c 18), and
(b) authorised by section 71 of that Act as purposes for which moneys may be raised by mortgage,

include the payment of any expenses incurred by a tenant for life or statutory owner as a member of a RTM company.

## 110 Power to prescribe procedure
(1) Where a claim to acquire the right to manage any premises is made by the giving of a claim notice, except as otherwise provided by this Chapter—

(a) the procedure for giving effect to the claim notice, and
(b) the rights and obligations of all parties in any matter arising in giving effect to the claim notice,

shall be such as may be prescribed by regulations made by the appropriate national authority.

(2) Regulations under this section may, in particular, make provision for a person to be discharged from performing any obligations arising out of a claim notice by reason of the default or delay of some other person.

## 111 Notices
(1) Any notice under this Chapter—

(a) must be in writing, and
(b) may be sent by post.

(2) A company which is a RTM company in relation to premises may give a notice under this Chapter to a person who is landlord under a lease of the whole or any part of the premises at the address specified in subsection (3) (but subject to subsection (4)).

(3) That address is—

(a) the address last furnished to a member of the RTM company as the landlord's address for service in accordance with section 48 of the 1987 Act (notification of address for service of notices on landlord), or
(b) if no such address has been so furnished, the address last furnished to such a member as the landlord's address in accordance with section 47 of the 1987 Act (landlord's name and address to be contained in demands for rent).

(4) But the RTM company may not give a notice under this Chapter to a person at the address specified in subsection (3) if it has been notified by him of a different address in England and Wales at which he wishes to be given any such notice.

(5) A company which is a RTM company in relation to premises may give a notice under this Chapter to a person who is the qualifying tenant of a flat contained in the premises at the flat unless it has been notified by the qualifying tenant of a different address in England and Wales at which he wishes to be given any such notice.

## Interpretation

### 112 Definitions

(1) In this Chapter—

"appurtenant property", in relation to a building or part of a building or a flat, means any garage, outhouse, garden, yard or appurtenances belonging to, or usually enjoyed with, the building or part or flat,

"copy", in relation to a document in which information is recorded, means anything onto which the information has been copied by whatever means and whether directly or indirectly,

"document" means anything in which information is recorded,

"dwelling" means a building or part of a building occupied or intended to be occupied as a separate dwelling,

"flat" means a separate set of premises (whether or not on the same floor)—

(a) which forms part of a building,
(b) which is constructed or adapted for use for the purposes of a dwelling, and
(c) either the whole or a material part of which lies above or below some other part of the building,

"relevant costs" has the meaning given by section 18 of the 1985 Act,

"service charge" has the meaning given by that section, and

"unit" means—

(a) a flat,
(b) any other separate set of premises which is constructed or adapted for use for the purposes of a dwelling, or
(c) a separate set of premises let, or intended for letting, on a tenancy to which Part 2 of the Landlord and Tenant Act 1954 (c 56) (business tenancies) applies.

(2) In this Chapter "lease" and "tenancy" have the same meaning and both expressions include (where the context permits)—

(a) a sub-lease or sub-tenancy, and
(b) an agreement for a lease or tenancy (or for a sub-lease or sub-tenancy),

but do not include a tenancy at will or at sufferance.

(3) The expressions "landlord" and "tenant", and references to letting, to the grant of a lease or to covenants or the terms of a lease, shall be construed accordingly.

(4) In this Chapter any reference (however expressed) to the lease held by the qualifying tenant of a flat is a reference to a lease held by him under which the demised premises consist of or include the flat (whether with or without one or more other flats).

(5) Where two or more persons jointly constitute either the landlord or the tenant or qualifying tenant in relation to a lease of a flat, any reference in this Chapter to the landlord or to the tenant or qualifying tenant is (unless the context otherwise requires) a reference to both or all of the persons who jointly constitute the landlord or the tenant or qualifying tenant, as the case may require.

(6) In the case of a lease which derives (in accordance with section 77(5)) from two or more separate leases, any reference in this Chapter to the date of the commencement of the term for which the lease was granted shall, if the terms of the separate leases commenced at different dates, have effect as references to the date of the commencement of the term of the lease with the earliest date of commencement.

## 113 Index of defined expressions

In this Chapter the expressions listed below are defined by the provisions specified.

*Appendix*

| Expression | Interpretation provision |
|---|---|
| Approval (and approving) | Section 98(7) |
| Appurtenant property | Section 112(1) |
| Acquisition date | Sections 74(1)(b) and 90 |
| Claim notice | Section 79(1) |
| Contractor party | Section 91(2)(b) |
| Copy | Section 112(1) |
| Counter-notice | Section 84(1) |
| Date of the commencement of the term of a lease | Section 112(6) |
| Determination date | Section 91(5) |
| Document | Section 112(1) |
| Dwelling | Section 112(1) |
| Existing management contract | Section 91(3) |
| Flat | Section 112(1) |
| Landlord | Section 112(3) and (5) |
| Lease | Section 112(2) to (4) |
| Letting | Section 112(3) |
| Long lease | Sections 76 and 77 |
| Manager party | Section 91(2)(a) |
| No dispute about entitlement | Section 90(3) |
| Notice of invitation to participate | Section 78 |
| Notice of withdrawal | Section 86(1) |
| Premises to which this Chapter applies | Section 72 (and Schedule 6) |
| Qualifying tenant | Sections 75 and 112(4) and (5) |
| Relevant costs | Section 112(1) |
| Relevant date | Section 79(1) |
| Right to manage | Section 71(2) |
| RTM company | Sections 71(1) and 73 |
| Service charge | Section 112(1) |
| Tenancy | Section 112(2) |
| Tenant | Section 112(3) and (5) |
| Tenant covenant | Section 100(4) |
| Unit | Section 112(1) |

©Crown Copyright

# Index

1967 Act ............................................. 2-08/2-17
   house ........................................... 2-12/2-17

1987 Act ............... 2-03/2-07,3-01/3-42, 4-01/4-491, 6-01/6-12
   acceptance notice ........................... 3-17/3-20, 4-133
      duly served ..................................... 4-134
   acceptance period
      section 5 notices ............................... 4-126
   affecting any premises ............................... 4-04
   agreement for lease .................................. 4-29
   application of ................................. 3-01/3-11
   appurtenance ........................................ 4-65
   appurtenant premises *see* appurtenance
   associated company ................................... 4-62
   auction *see* sale at a public auction
   avoidance ...................................... 6-01/6-12
      assignment of rights under contract ................. 4-74
      associated company, use of ......................... 6-09
      contract, held in escrow ........................... 6-10
      disposal, take outside the Act ................. 6-09/6-11
      election, disposal for non-monetary consideration .... 4-204
      landlord, take outside the Act ................. 6-05/6-07
      lease, intervening *see* landlord, take outside the Act
      mortgage .......................................... 6-11
      premises, take outside the Act ..................... 6-04
      qualifying tenants, limit number .................... 6-08
      share sale ........................................ 6-12

best endeavours ................................... 4-213
body corporate *see* associated company
building ..................................... 3-07, 4-17
charity ........................................... 4-60
commercial unit .................................. 4-47
common parts .................................... 4-50
consent, third party ......................... 4-213/4-224
constituent flats
    section 5 ..................................... 4-100
    section 11 .................................... 4-309
contract .......................................... 4-47
    auction ...................................... 4-194
    conditions, time for fulfillment by nominated
        person .............................. 4-335, 4-337
    entering into ................................. 4-320
    entering into, nominated person deemed to have ..... 4-338
costs ...................................... 4-394/4-395
    joint and several liability ..................... 4-245
    nominated person, liability *see* withdrawal
        (nominated person)
court ............................................ 4-450
    enforcement of obligations ................ 4-482/4-487
criminal offence ........................... 4-282/4-292
    fine .......................................... 4-286
    local housing authority ........................ 4-290
    without reasonable excuse ..................... 4-283
Crown ...................................... 3-08, 4-29
deposit .......................................... 4-334
disposal
    by way of security for loan ..................... 4-58
    subsequent, in breach of Act ................... 4-210
duly nominated ............................ 4-137, 4-140
dwelling ......................................... 4-19
exceeds 50% ..................................... 4-21
exempt landlord *see* landlord, exempt
flat ........................................ 3-06, 4-19
    single flat *see* Tenants' rights, single flat
incorporeal hereditament .......................... 4-57
incumbrance ..................................... 4-362
interested person ................................. 4-484

# Index

landlord
    exempt ................................... 3-08, 4-29
    Crown as .................................. 3-08, 4-29
    immediate ...................................... 4-34
    resident .................................. 3-08, 4-29
landlord's rights after acceptance and nomination ..... 3-21–3-34
lapse of landlord's offer .............. 3-33 - 3-34, 4-267/ 4-280,
lease *see* tenancy
Leasehold Valuation Tribunal
    application to ................................. 4-208
    jurisdiction of ........................... 4-397/4-399
    nominated person, role *see* representing tenants
        in proceedings
local housing authority ............................ 4-290
member of family ............................... 4-69/4-70
mixed use
    building ......................................... 3-05
    history ...................................... 1-01/1-06
mortgagee ......................................... 4-29
    following provisions ............................ 4-55
    sale by ......................................... 4-54
nominated purchaser .................... 4-130, 4-388/4-393
    more than one ................................. 4-144
    replacement .................................... 4-142
    representing tenants in proceedings .......... 4-400/4-401
    support by tenants ........................ 4-130, 4-140
notice ............................................. 4-83
    prospective purchaser *see* section 18 notice
notice of withdrawal (landlord) .4-250, 4-172, 4-174, 4-250/4-265
    costs .......................................... 4-260
notice of withdrawal (nominated person) .......... 4-226/4-248
    costs ..................................... 4-241/4-245
notice under section 12B
    service ........................................ 4-349
notice under section 12C
    service ........................................ 4-376
notice seeking information about disposal ...................
    form .......................................... 4-307
    service ........................................ 4-307
obliged to proceed ........................... 4-178, 4-184

offer, by landlord
    section 5A ........................................4-101
    section 5B ........................................4-107
    section 5C ........................................4-114
offer notice *see* offer, by landlord
person or persons nominated by the tenants
    *see* nominated purchaser
premises ...............................3-04/3-07, 4-05
    affected at time of disposal ........................4-06
previously served ......................................4-08
principal terms of contract *see* principal terms of disposal
principal terms of disposal ...........................4-475
    section 5A .........................................4-97
    section 5B ........................................4-105
    section 5C ........................................4-112
    section 5D ........................................4-117
proceeding with disposal
    disposal at public auction ................3-29/3-30, 4-186
    disposal not at public auction .............3-26/3-28, 4-177
proposes to make a relevant disposal ....................4-79
protected interest .....................................4-129
protected period ......................................4-135
purchaser ............................................4-312
    rights against landlord ............................4-325
    section 11 ........................................4-312
    section 18 ........................................4-466
qualification ......................3-01/3-11, 3-38, 4-01/4-77
qualifying tenant ............................3-09/3-10, 4-37
relevant disposal .................................3-11, 4-47
requirements, sections 6–10 ............................. 4-15
requisite majority of qualifying tenants ...3-18, 4-98, 4-480/4-481
    section 5 .........................................4-99
    section 11A ..............................4-308, 4-316
    section 12A, 12B & 12C ...........................4-321
resident landlord *see* landlord, resident
residential purposes ................................... 4-23
restriction on disposal, landlord ........4-124, 4-132, 4-160, 4-257
    ending of restrictions ............4-146, 4-154, 4-156, 4-158,
    ...................................4-237/4-239, 4-280
right to information
    conferred ........................................4-14

## Index

rights of first refusal ................................. 4-13
sale at a public auction .............................. 4-104
    rights of nominated person, revised timetable ..4-199/4-200
    rights of successful bidder ....................... 4-197
Secretary of State
    order by ......................................... 4-27
    regulations by .................................. 4-491
section 5 notice ...................................... 4-81
section 5 requirements .......................... 3-12/3-16
    compliance ................................. 4-88, 4-94
    section 5A ............................... 4-96/4-101
    section 5B .............................. 4-103/4-109
    section 5C .............................. 4-111/4-114
    section 5D .............................. 4-116/4-118
    section 5E .............................. 4-120/4-122
section 5, tenants' response ..................... 3-17/3-20
section 18 notice ............................. 4-460/4-479
serve
    acceptance notice, duly served .................... 4-134
    different dates ................................... 4-92
    notice, general .................................. 4-82
    notice giving information of disposal .......... 4-307, 4-316
    section 18 notices ......................... 4-476/4-479
sever ........................................ 3-13, 4-0
statutory tenancy .................................... 4-32
statutory tenant *see* statutory tenancy
subsequent purchaser
    tenants' rights against ..................... 4-418/4-435
tenancy ..................................... 4-29, 4-39
tenant ..................................... 4-36, 4-490
tenants' rights
    auction, after sale at public ...................... 4-196
    compel sale etc by purchaser ................ 4-341/4-367
    contract, take benefit ...................... 4-317/4-340
    increase in value of property ............... 4-366/4-367
    incumbered property ....................... 4-359/4-365
    information about purchase, receipt of ........ 4-306/4-316
    new lease, compel grant of ................. 4-368/4-387
    non-service of notice, effect ...................... 4-94
    notice seeking disposal information *see* notice
        seeking information about disposal

   option/right of pre-emption, in relation to . . . . . . . . . . .4-320
   property, within and without the Act . . . . .4-340, 4-357, 4-386
   purchaser, against . . . . .3-35/3-37, 4-293/4-459, 4-388/4-393,
   . . . . . . . . . . . . . . . . . . . . . . . . . . . . . . . . . . . . . . . . . . . .4-396
   subsequent purchaser, against *see* subsequent purchaser
   termination of . . . . . . . . . . . . . . . . . . . . . . . . . . . .4-436/4-459
   time limits for exercise . . . . . . . . . . .4-307, 4-314, 4-315, 4-329,
   . . . . . . . . . . . . . . . . . . . . . . . . . . .4-352, 4-380, 4-448/4-459
  transferee . . . . . . . . . . . . . . . . . . . . . . . . . . . . . . . . . . . . . . . . . .4-488
  withdrawal (landlord) . . . . . . . . . . . . . . . . . .3-22, 3-23/3-24, 4-180
  withdrawal (nominated person) . . . . . . . . .3-22, 3-31- 3-32, 4-182,
  . . . . . . . . . . . . . . . . . . . . . . . . . . . . .4-201/4-202, 4-402/4-417
   costs liability . . . . . . . . . . . . . . . . . . . . . . . . . . . .4-413/4-4-417

1993 Act . . . . . . . . . . . . . . . . . . . . . . . . . . . . . . . . . .2-18/2-20, 5-01/5-60
  avoidance . . . . . . . . . . . . . . . . . . . . . . . . . . . . . . . . . . . . .6-13/6-15
   premises, taking outside Act . . . . . . . . . . . . . . . . . . . . . . .6-15
   qualifying tenants, limiting number . . . . . . . . . . . . . . . . . .6-15
  estate management scheme . . . . . . . . . . . . . . . . . . . . . . . .6-16/6-21
  legislation . . . . . . . . . . . . . . . . . . . . . . . . . . . . . . . . . . . . . . . . .5-05
  long lease . . . . . . . . . . . . . . . . . . . . . . . . . . . . . . . . . . . . . . . . . 5-16
  premises . . . . . . . . . . . . . . . . . . . . . . . . . . . . . . . . . . . . . 5-09/5-15
  price *see* procedure, valuation
   procedure . . . . . . . . . . . . . . . . . . . . . . . . . . . . . . . . . .5-23/5-50
   conveyance . . . . . . . . . . . . . . . . . . . . . . . . . . . . . . . . .5-51/5-55
   counter-notice . . . . . . . . . . . . . . . . . . . . . . . . . . . . . . .5-34/5-40
   disputes . . . . . . . . . . . . . . . . . . . . . . . . . . . . . . . . . . .5-45/5-50
   initial notice . . . . . . . . . . . . . . . . . . . . . . . . . . . . . . . .5-26/5-29
   initial notice, effect of . . . . . . . . . . . . . . . . . . . . . . . . .5-30/5-33
   landlord's notice *see* counter-notice
   leaseback . . . . . . . . . . . . . . . . . . . . . . . . . . . . . . . . . . .5-41/5-44
   reversioner, ascertaining . . . . . . . . . . . . . . . . . . . . . . . . . . .5-25
   tenants' notice *see* initial notice
   timetable . . . . . . . . . . . . . . . . . . . . . . . . . . . . . . . . . . .5-24, 5-60
   valuation . . . . . . . . . . . . . . . . . . . . . . . . . . . . . . . . . . .5-56/5-59
  qualification . . . . . . . . . . . . . . . . . . . . . . . . . . . . . . . . . . .5-06/5-20
  qualifying tenant . . . . . . . . . . . . . . . . . . . . . . . . . . . . . . . . 5-16/5-20
  right to enfranchise company *see* RTE company
  RTE company . . . . . . . . . . . . . . . . . . . . . . . . . . . . . . . . . . . . . . .5-22
  tenants' rights
   qualification . . . . . . . . . . . . . . . . . . . . . . . . . . . . . . .5-06/5-5-20

*Index*

withdrawal
    tenants, deemed .................................5-21

2002 Act ................................2-21/2-23, 5-61/5-84
    acquisition date ........................................5-79
        premises, taking outside Act .......................6-15
        qualifying tenants, limiting number .................6-15
    avoidance ..................................... 6-13/6-15
    landlord problem areas ...............................5-83
    legislation ..........................................5-63
    procedure .....................................5-65/5-81
        contractor notices ................................5-80
        landlord's counter-notice .....................5-75/5-76
        notice of claim ..............................5-69/5-70
        notice of claim, effect of ......................5-71/5-74
        notice of invitation to participate ...............5-67/5-68
        request for information ......................5-77/5-78
        service charge/sinking/reserve funds ................5-81
        tenants' notice *see* notice of claim
        timetable ........................................5-84
    right to manage company *see* RTM company
    RTM company .......................................5-66
        landlord's membership ............................5-82
    qualification .........................................5-64

Collective enfranchisement (flats) *see* 1993 Act
Commonhold and Leasehold Reform Act 2002 *see* 2002 Act

Enfranchisement (houses) *see* 1967 Act

Landlord and Tenant Act 1987 *see* 1987 Act
Leasehold Reform Act 1967 *see* 1967 Act
Leasehold Reform, Housing and Urban Development Act 1993
    *see* 1993 Act

Right to Manage *see* 2002 Act

Tenants' rights of first refusal *see* 1987 Act